CUBAN BLINDNESS

CUBAN BLINDNESS
Diary of a Mysterious Epidemic Neuropathy

GUSTAVO C. ROMÁN

Amsterdam • Boston • Heidelberg • London
New York • Oxford • Paris • San Diego
San Francisco • Singapore • Sydney • Tokyo
Academic Press is an imprint of Elsevier

Academic Press is an imprint of Elsevier
125, London Wall, EC2Y 5AS, UK.
525 B Street, Suite 1800, San Diego, CA 92101-4495, USA
225 Wyman Street, Waltham, MA 02451, USA
The Boulevard, Langford Lane, Kidlington, Oxford OX5 1GB, UK

Notices

Knowledge and best practice in this field are constantly changing. As new research and
experience broaden our understanding, changes in research methods, professional practices,
or medical treatment may become necessary.

Practitioners and researchers must always rely on their own experience and knowledge
in evaluating and using any information, methods, compounds, or experiments described
herein. In using such information or methods they should be mindful of their own safety
and the safety of others, including parties for whom they have a professional responsibility.

To the fullest extent of the law, neither the Publisher nor the authors, contributors, or
editors, assume any liability for any injury and/or damage to persons or property as a
matter of products liability, negligence or otherwise, or from any use or operation of any
methods, products, instructions, or ideas contained in the material herein.

ISBN: 978-0-12-804083-6

British Library Cataloguing-in-Publication Data
A catalogue record for this book is available from the British Library.

Library of Congress Cataloging-in-Publication Data
A catalog record for this book is available from the Library of Congress.

For Information on all Academic Press publications
visit our website at http://store.elsevier.com

Typeset by MPS Limited, Chennai, India
www.adi-mps.com

Printed and bound in the United States of America.

CONTENTS

ABOUT THE AUTHOR

Gustavo C. Román is a clinical neurologist practicing in Houston, Texas. During his academic career, he has been interested in neuroepidemiology, tropical neurology, stroke, and vascular, nutritional, and infectious diseases of the nervous system. Most recently, he has studied autism and Alzheimer disease. Since 2010, he works at the Neurological Institute of Houston Methodist Hospital, where he is the Jack S. Blanton Presidential Distinguished Endowed Chair for the Study of Neurological Disease and Professor of Neurology at Weill Cornell Medical College, Cornell University, New York.

ACKNOWLEDGMENTS

It always seem impossible until it's done
–Nelson Mandela

Many people provided support, information, and help during the writing of this book. I want to express my gratitude to my colleagues at the National Institute of Health in Washington, DC, including those cited in the book as well as those not mentioned. Without the support of the World Health Organization (WHO) and the Pan-American Health Organization (PAHO), the Mission to Cuba would not have been possible. Fieldwork in Cuba with the members of the Mission to Cuba team created links of friendship and camaraderie that have endured to this day. This group includes members of the nonprofit ophthalmology group Orbis as well as colleagues from the Centers for Disease Control and Prevention (CDC), the U.S. Food and Drug Administration (FDA), and Emory University in Atlanta.

In Cuba, many people offered warm hospitality to the Mission, along with abundant epidemiologic and clinical information. I want to acknowledge the logistic help provided by many members of the Cuban government, in particular those at the Ministry of Health. Numerous epidemiologists, ophthalmologists, neurologists, internists, basic scientists, and pathologists in Havana, Pinar del Río, Santiago, and Youth Island openly shared with us information on many aspects of the epidemic. The late Dr Miguel Marquez, PAHO representative in Havana, and his wife Libia Cerezo opened their home to us generously. At the Cuban Institute of Neurology and Neurosurgery, Drs Ricardo Santiago-Luis González and Rosaralis Santiesteban provided friendship and invaluable clinical and historical information.

I want to acknowledge the support of Charles Fleming, author and editor *par excellence*, who helped me transform the initial manuscript into the current book. At Elsevier, Peter Frans Bakker, Natalie Farra, Kristi L. Anderson, and the Academic Press editorial group ensured a swift publication process. Also, I am grateful to Prous Science, Barcelona, Spain, for their long-term support.

Last but not least, I want to dedicate this book to my family, in particular to my wife Dr Lydia Navarro Román, who provided endless support, especially for my research travel to Cuba, and her insightful comments on the manuscript; to my sons Gustavo and Andrés, who helped me with resolving grammatical issues and providing style comments; and, of course, to my daughter Natalia Isabel Román Navarro *in memoriam.*

Gustavo Román
Houston 2015

PROLOGUE

Comrades! This is the most difficult period in Cuba's history.

Fidel Castro had taken the podium to make his final remarks at the closing session of the National Assembly at Havana's Palace of Conventions.[1] It was December 27, 1991. His nation was facing a crisis. The Soviet Union, Cuba's principal economic sponsor, was in a state of collapse. The United States, Cuba's principal economic foe, had enacted new legislation, prohibiting all US companies from conducting any trade with the Caribbean nation.[2] No vessel carrying Cuban passengers or Cuban goods would be allowed to enter any US port, nor would any vessel delivering goods to Cuba be allowed to enter for 6 months after departing from a Cuban seaport. The US government enforced these new restrictions under its "Trading with the Enemy Act." The blockade of Cuba was about to become law.

Castro understood the implications of these new developments. The year ahead, 1992, he said, would be crushing to the Cuban people.

Comrades, be prepared for the worst circumstances …

Date: Monday, 12/30/1991 Bethesda, Maryland: I am closing my second year as Chief of the Neuroepidemiology Branch at the NIH.[3] Although the word "Neuroepidemiology" on my door implies that we deal primarily with epidemic diseases affecting the nervous system, in reality, large epidemics of neurologic diseases are more a thing of the past. Vaccination, environmental sanitation, and public health improvements have controlled polio and meningococcal meningitis almost completely. The closest instances of epidemics would be the small clusters of tropical spastic paraparesis (TSP), which we have studied in Kagoshima (Japan), Tumaco (Colombia), Ecuador, Brazil, the Caribbean islands, and the Seychelles in the Indian Ocean. This is an interesting disease, which showed the neurologic effects of the first-recognized human retrovirus, referred to as HTLV-I.[4] This infectious agent is capable of also causing lymphoma and leukemia in humans and animals; it belongs to the same retrovirus family as HIV, which causes AIDS.[5]

The one achievement this year that will have long-lasting benefits was the international meeting of experts to define the research criteria for vascular dementia, held from April 19–21, 1991, on the NIH campus here in Bethesda. I organized it with sponsorship from the NINDS[6] and travel support from AIREN,[7] the nonprofit foundation of Professor Liana Bolis in Geneva. The NINDS-AIREN criteria will be used for future research in this field, including controlled clinical trials to treat patients with vascular dementia.[8] Fortunately, all seems to be quiet at the year's end in the snow-covered fields of the NIH campus. Let's hope for a peaceful and productive 1992.

ENDNOTES

1. *Speech by President Fidel Castro at the closing session of the National Assembly of the People's Government at the Palace of Conventions in Havana on December 27, 1991.* **Source Line: PA3112184591 Havana Radio and Television Networks in Spanish. 2300 GMT 29 Dec 91 [Abbreviated Text] www.LANIC.utexas.edu/la/cb/cuba/castro.html**.

    Comrades, worry not because I plan to be brief.

    This is the most difficult period in Cuba's history. It is not just the most difficult period of the revolution, but the most difficult in Cuba's history. I call on each one of you to first think about this fact. The most difficult year is the one that begins now in 1992. I could not call it the year of the "special period" because the "special period" will last more than a year.

    Be prepared for the worst circumstances. If 7 million tons of fuel are needed, I say let us plan on four. The enemy is busy trying to make 1992 a difficult year for us, as difficult as possible. It will try to prevent us from having any market for our products, from getting fuel, not even for cash. The enemy is busy in Moscow trying to hurt our economic ties with whatever is left of the USSR. They are trying to block, in every way possible, our effort to get at least one third of the fuel we need to make ends meet. We know how the enemy works to make the blockade more effective and to make our lives more difficult. This is why we must always calculate based on the worst circumstances. The worst is about to come in 1992.

    To all this we must add the disappearance of trade with the socialist sphere. It is a very sad reality, but it is a reality. The USSR does not exist. I said two-and-a-half years ago that if one day we woke up to the news that the USSR no longer existed we would continue to defend the revolution and socialism. The revolution and the country were going to be left in a difficult, very difficult situation. But we had to do exactly what we are doing; enter the "special period in peacetime," implement the appropriate measures, prepare ourselves for an even more difficult phase, conscious that we are reaching the limit of the damage that can be done to us.

    I asked the comrades if they were ready and prepared to face the most complicated and critical "special period," if they can feed the population, and if they can solve all these problems. [For transportation] almost 500,000 bicycles have been distributed. But in addition, there are dozens and dozens of agricultural camps, thousands and thousands of people mobilized, 32 contingents, an entire planting program.

    Of course, our standards are the most just. We do not leave a single citizen unprotected. We do not leave a single citizen, university graduate or middle-level technician who graduates, to his fate. We do not leave anyone to their fate. A capitalist country could not do that. Only a socialist system could do what we are doing, with this principle of distributing what we have among all of us, as in a family.

There are many ideas and we are working in many directions to find solutions to the problems. Now, we will unquestionably have to undergo limitations. This is inevitable. We have to go through a period of sacrifices. You can see how up to now we have been able to keep all schools and hospitals in operation. The teachers have shown extraordinary willingness to maintain education under any circumstances in the "special period." The scientists, technicians, skilled workers, engineers, everyone, are willing. They presented 34,000 papers, 40,000 solutions at the forum. This is what our country today is capable of doing.

This collapse has been very quick, very rapid. As we have said, if some of the things that have occurred had taken 4 or 5 years to happen, how many problems would we have solved? But it happened too fast, and it has created some imbalances in our programs. We must be prepared for imbalances. You, who lead the people, and lead the party members, and lead the mass organizations, must prepare them for all these contingencies and especially for the more critical situations that will arise.

I say and I repeat that it is more important to have brains as a natural resource than to have oil. With that resource, which is our people's intelligence, which we have planted in their hearts and minds for 30 years, we will move forward. We will move forward, there is no doubt, first slowly and then faster.

That is why I am expressing here my conviction that this generation of Cubans, and those who come afterwards, and those who came before us, will be respected and admired. They will always have to recognize that we were able to fight in the most incredibly difficult conditions, that we were willing to give everything to defend everything that we hold sacred, that we were able to give everything for victory. Because if we act as we should act, if we conduct ourselves as we should conduct ourselves, and as I am sure we will, victory is the only possible end result. [Applause]

Socialism or death, fatherland or death, we will win. [Applause]

Let us go to work now. I wish you a heroic, glorious, and successful 1992. [Applause]

2. *The Cuban Democracy Act of 1992*: Congressman Robert G. Torricelli, New Jersey's Democratic representative, and Democratic senator for Florida Bob Graham, presented to the 102 US Congress a bill called *The Cuban Democracy Act* that was signed into law in October 1992 by president George Bush Sr, then in the midst of a reelection campaign. Endorsement of the law by the eventual winner of the presidential election, Bill Clinton, received the enthusiastic approval of the predominantly anti-Castro Cuban-American community in Florida.

The bill was expected "to promote a peaceful transition to democracy in Cuba through the application of sanctions directed at the Castro government and support for the Cuban people." The law was based on the expected impact on Cuba's government of the political changes in Western Europe that led to the disappearance of Communist governments; it listed a number of findings by Congress, including among others the following:

Events in the former Soviet Union and Eastern Europe have dramatically reduced Cuba's external support and threaten Cuba's food and oil supplies.

The fall of communism in the former Soviet Union and Eastern Europe, the now universal recognition in Latin America and the Caribbean that Cuba provides a failed model of government and development, and the evident inability of Cuba's economy to survive current trends, provide the United States and the international democratic community with an unprecedented opportunity to promote a peaceful transition to democracy in Cuba.

However, Castro's intransigence increases the likelihood that there could be a collapse of the Cuban economy, social upheaval, or widespread suffering. The recently concluded Cuban Communist Party Congress has underscored Castro's unwillingness to respond positively to increasing pressures for reform either from within the party or without.

The United States cooperated with its European and other allies to assist the difficult transitions from Communist regimes in Eastern Europe. Therefore, it is appropriate for those allies to cooperate with United States policy to promote a peaceful transition in Cuba.

Based on the above, the law formulated 10 policy points, including the following two: It should be the policy of the United States—

1. To seek a peaceful transition to democracy and a resumption of economic growth in Cuba through the careful application of sanctions directed at the Castro government and support for the Cuban people.
10. To initiate immediately the development of a comprehensive United States policy toward Cuba in a post-Castro era.

The law explicitly prohibited foreign-based subsidiaries of US companies from trading with Cuba, prevented travel to Cuba by US citizens, and prohibited family remittances to Cuba. Specifically, the law discouraged all nations to conduct commerce with Cuba:

The President should encourage the governments of countries that conduct trade with Cuba to restrict their trade and credit relations with Cuba in a manner consistent with the purposes of this title.

A vessel, which enters a port or place in Cuba to engage in the trade of goods or services, may not, within 180 days after departure from such port or place in Cuba, load or unload any freight at any place in the United States.

A vessel carrying goods or passengers to or from Cuba or carrying goods in which Cuba or a Cuban national has any interest may not enter a United States port.

The President shall establish strict limits on remittances to Cuba by United States persons for the purpose of financing the travel of Cubans to the United States, in order to ensure that such remittances reflect only the reasonable costs associated with such travel, and are not used by the government of Cuba as a means of gaining access to United States currency.

Finally, the law was enforced by the severe sanctions imposed by the Trading with the Enemy Act:

The authority to enforce this title shall be carried out by the Secretary of the Treasury. The Secretary of the Treasury shall exercise the authorities of the Trading With the Enemy Act in enforcing this title.

Source: Cuban Democracy Act (CDA) www.treasury.gov/resource-center/sanctions/ Documents/cda.pdf.

3. NIH = National Institutes of Health, Department of Health and Human Services, United States of America.
4. HTLV-1 = Human T-lymphotropic virus, type 1.
5. HIV = Human immunodeficiency virus; AIDS = acquired immunodeficiency syndrome.

6. NINDS = National Institutes of Neurological Disorders and Stroke, National Institutes of Health, Department of Health and Human Services, United States of America.
7. AIREN = Association Internationale pour la Recherche et l'Enseignement en Neurosciences, Geneva, Switzerland.
8. Román GC, Tatemichi TK, Erkinjuntti T, Cummings JL, Masdeu JC, García JH, Amaducci L, Orgogozo JM, Brun A, Hofman A, Moody DM, O'Brien MD, Yamaguchi T, Grafman J, Drayer BP, Bennett DA, Fisher M, Ogata J, Kokmen E, Bermejo F, Wolf PA, Gorelick PB, Bick KL, Pajeau AK, Bell MA, DeCarli C, Culebras A, Korczyn AD, Bogousslavsky J, Hartmann A, and Scheinberg P. Vascular Dementia: Diagnostic Criteria for Research Studies. Report of the NINDS-AIREN International Workshop. *Neurology* 1993;43(2):250–260.

CHAPTER 1

Pinar del Río
Cuba, December 12, 1991

An eye for an eye only leads to more blindness.
Margaret Atwood

The westernmost end of the island of Cuba, the province of Pinar del Río, produces the finest tobacco in the world. The names of the municipalities of Candelaria, San Cristobal, San Diego de los Baños, and Consolación del Sur deep inside Cuba's tobacco lands are pronounced with utmost respect by cigar connoisseurs worldwide.[1] The greenhouse microclimate of the tobacco valleys, or *vegas*,[2] the relatively poor soil and the time-honored tradition of "resting the soil" between harvests combined with careful selection of seedlings all contribute to the uniformly superb quality of these tobaccos. Cultivation is laborious: the small plots of land are tilled with yokes of oxen, and the seedlings are planted by hand. During the first 6 months of life, the plants are covered with white muslin to protect them from the tropical sun, giving the appearance of windblown sails floating over the ocher fields.

For generations, tobacco has been the life and work of this region. Enjoying the soft perfumed smoke of a *Cohiba* Corona—the world's finest cigar—no one can imagine the intensity and demands of the labor required for the production of each cigar. About 80% of Cuba's tobacco is produced by fiercely independent farmers, or *guajiros*,[3] cultivating small patches of land in Pinar. The work is slow, arduous, and very labor intensive. This explains the high population density of municipalities such as San Juan y Martínez, San Luis, Consolación del Sur, Güane, and Sandino. Residents of these cobblestone-street villages crowd into diminutive red-tile houses, each with its front porch and small patio gardens, spread along the main roads. These villages reflect the traditions of the Spanish colonists who left the Canary Islands to settle here. It was in these densely populated but tranquil lands that the epidemic of blindness exploded.

José Polo Portilla,[4] a true *guajiro*, had spent his entire adult life working a patch of tobacco field in the village of San Juan y Martínez, near the city

Cuban Blindness. DOI: http://dx.doi.org/10.1016/B978-0-12-804083-6.00001-1

of Pinar del Río. He was a widower living alone. His only daughter, married 2 years earlier, visited him nearly every week. During idle time away from the fields, he would roll his own tobacco into thick cigars, which he would enjoy smoking while listening to the radio. News of the shortage of the food available through rationing cards did not affect him greatly. He was a lean, worn man of 57 years of age, with sun-wrinkled skin that made him look much older. He had always had a small appetite and was satisfied with his daily portion of rice and black beans. He also managed to grow *yuca*, the Cuban cassava, in his garden. But he saved the few eggs allowed him by his government ration for the youngest of his grandchildren. Like most Cuban men of his age, he enjoyed a bottle of white rum, which also was part of his government ration, as it had been for almost 30 years since the days of the triumph of the Revolution.

Not having coffee, however, was difficult. Something he just could not get used to. José grew up drinking the thick, aromatic beverage, brewed in the Arab fashion—a custom deeply anchored in Spain after seven centuries of Arab domination.[5] Every day, sometimes several times a day, he would savor a small cup of dark-roasted coffee—freshly boiled, strained, and made very sweet with two heaping spoons of sugar. There was nothing better when he arrived home in the midsummer heat, dripping with sweat, than to quench the thirst with an almost-boiling *taza de café* from Santiago de Cuba followed by a tall glass of fresh water. But now coffee was so scarce that he was drinking what was practically sugary hot water for breakfast and throughout the day.

In the last week of November 1991, José noticed that the bright sunlight bothered his eyes when he went out to the fields to check his plants for the coming harvest. He resorted to using an old wide-brimmed straw hat, tilted over his face, to shade his eyes. He began to have difficult reading, even with the glasses that had helped in the past. Also, he was having trouble sleeping because he was waking up at night several times to urinate. He had begun to feel unusually tired, too, and would take an unscheduled *siesta* at the end of the day. His daughter noticed that he looked thinner, and to please her he promised to go to his doctor—after the tobacco season ended.

This year's harvest was unpromising. The lack of rains and governmental experiments with new untried fertilizers and antifungals to control the "blue mold" had produced a tobacco crop of disappointing quality.[6] José Polo's health was worsening, too. By mid-December, his eyesight had deteriorated so much that it felt as if he were looking through fogged lenses, and he asked his barber to remove his white mustache: He could no longer trim it properly. He had also failed to notice this year the spectacular orange-red flowering of the Royal Poinciana trees, the Cuban *Flamboyant* trees in his backyard. He was also getting quite hard of hearing, and one night he was awakened by the high-pitched sound of a cicada inside his room. He searched for the insect until he realized that the sound was inside his ears beeping constantly in the silence of the night.

"Old age never arrives by itself," he thought and decided to go with his daughter to the nearby *dispensario*.[7] A young doctor examined him thoroughly and found that José could hardly see the big letters on the screen and sent him to the eye specialist in the regional hospital in Pinar del Río.

The Regional Hospital "Abel Santamaría" is a modern four-storey building situated on a small hill at the entrance of Pinar del Río off the highway from Havana. It has long corridors surrounding a central courtyard filled with flowers, palms, and trees providing shade and cool freshness in the tropical heat. Beside the hospital are a few blue buildings housing the schools of medicine and nursing. Young men and women in white coats move hurriedly across the central plaza to escape from the bright sun.

José now had to walk holding his daughter's arm because he could not see the floor in front of him. In just a few days his vision had clouded to the point where he could no longer recognize faces. José realized he was going blind.

After the fierce sunlight of the morning, the shaded examining room of the outpatient ophthalmology service felt refreshingly cool. Dr Blanca Emilia Elliot greeted him, and he removed his straw hat in deference. His examination took a long time. Dr Elliot showed him some books with dots, lines, and big letters. She had him sit facing a large black screen. White dots appeared and disappeared in front of him. Finally, she asked him to bend his head backward, and she put some drops in his eyes, which made his eyes itch and water. "This is to dilate your pupils," she explained.

She left him alone in the dark for several minutes, and then she returned to examine his eyes briefly and apply more drops. José kept his eyes tightly shut as the tears ran down his cheeks and inside his nose. He had never cried—not even when he found Caridad, his wife of 36 years, dead on the porch—but now the certainty of his blindness overwhelmed him with sadness and he wanted to weep.

The arrival of Dr Elliot interrupted his thoughts. She asked him to look at a small light while she observed his eyes through a long lens. Then she asked him, "How much do you smoke?"

"Well, three or four cigars a day," he answered.

"Since when?"

"Since I was 12, maybe."

"And how much do you drink?"

"Not much—one bottle of rum from the rations every week," he said defensively.

The doctor stood up, went outside, and returned with another doctor.

"This is Dr Carlos Perea, Chief of the Ophthalmology Service," she said. "I would like him to take a look at your eyes."

Dr Perea's hands had the fresh smell of soap. He looked carefully at José's eyes and then asked, "Have you noticed any problem with seeing the colors of objects?"

"Yes, doctor," José said. "It's so strange, but this year the red Flamboyant flowers don't look the same."

Dr Perea turned away from José Polo and said to Dr Elliot, "You're right. This is case number six of tobacco-alcohol amblyopia in 1 week."[8]

ENDNOTES

1. Gérard Père et Fils. *Havana cigars* (New York, 1997); M. Frank. Cigars 101. *Cigar Aficionado* (Autumn 1993); J. Suckling. Cuba's best cigars. *Cigar Aficionado* (June 1999).
2. Vegas (In Spain and Spanish America) = a large grassy plain or valley (Oxford dictionary).
3. Guajiro = peasant (Oxford dictionary).
4. Real names are not used.
5. L.P. Harvey. *Islamic Spain, 1250 to 1500* (Chicago, 1990).
6. Business notes: Tobacco, the Cuban stogie crisis. *Time Magazine* (9 April 1990).
7. Dispensario = dispensary, outpatient clinic.
8. Amblyopia = impaired or dim vision, blindness (Oxford dictionary); tobacco-alcohol amblyopia: visual impairment due to tobacco and alcohol abuse usually associated with nutritional deficiencies.

CHAPTER 2

An Economy in Crisis
Havana, November 22, 1991

The number one job is survival!
Fidel Castro

The highway from Havana to Pinar del Río was empty. The wide concrete roadway, built 30 years ago by the Soviets with military purposes in mind, had withstood the tropical sun, rains, hurricanes, and time. Now it was obsolete and ravaged by poverty. A large crowd waited interminably inside the fenced bus terminal by the roadside. Women squatted with babies on their laps protected by the shadow of white squares of cotton fabric held tight by their fingertips. Men in straw hats talked and smoked under the sun.

Derelict buses filled the parking lot. The Soviet Union had disappeared and with it the constant line of tanker ships bringing crude oil to the refinery of Regla across Havana Bay. Without Russian oil there had been no electricity and no fuel. No electricity meant no refrigeration, no air conditioning, and no industrial production. No fuel meant a gradual paralysis of all traffic. Trucks, buses, motorcycles, and automobiles had stopped, along with tractors and trains. Agricultural activities had ceased. Feed for cattle, pigs, and chickens had become unavailable. Without transportation food supplies had stopped. Cubans were left without rice, maize, plantains, cassava, yam, beans, manioc, tomatoes, and other produce; there were no seeds, fertilizers, or pesticides. To make matters worse, food reserves for the population were found to be at an all-time low. Without crude oil, Cuba could not earn the annual $500 million from selling gasoline and refined oil. Lost revenues meant no purchases of food and basic products in the world markets.

On November 22, 1991, speaking before the Fifth Congress of the Syndicate of Agriculture and Wildlife Workers, Fidel Castro summarized in a single lapidarian sentence the gravity of the situation: "The number one job is survival!"[1]

During the next few days the government recommended all Cubans to plant their own vegetable gardens on every available parcel of land. "Some

Cuban Blindness. DOI: http://dx.doi.org/10.1016/B978-0-12-804083-6.00002-3

27,000 gardens have been improvised in Havana," informed Ramón Estévez, provincial director of the Ministry of Construction.[2] The government also distributed free chicks to provide Cubans with a future source of eggs and much-needed animal protein.

Cuba's economic problems, however, had started much earlier. The Cuban economy had been disastrous since the early days of the Revolution when Ché Guevara signed the Cuban pesos. In 1962 Cubans had been forced into a state of almost constant rationing, limited distribution of items, and restrictions—all in the name of the common good. By 1976 about 80% of Cuban exports went to the socialist countries that were members of the Council on Mutual Economic Assistance (CMEA).[3] For the past 20 years the island had survived, thanks to the generous economic aid and subsidies from the Soviet Union, estimated at as much as $6 billion per year.[4]

Of course, a large percentage of this figure was represented by military items, including supplies, arms, canned food, meat, condensed milk, and dairy products for members of the Cuban Armed Forces. With a standing army of almost 200,000 soldiers plus another 110,000 reserve forces, Cuba was, by some accounts, the most militarized country in the world per capita.[5] The largest Cuban military expenses occurred in Africa, in particular, during the Internationalist Mission of Cubans in Angola—the so-called MICA (Misión Internacionalista de Cubanos en Angola)—which from 1976 until 1991 involved about 300,000 civilian and military personnel from Cuba (or 6% of the adult population of the island).[6]

In January 1986 Cuba was on the verge of bankruptcy, and the government implemented a series of "austerity measures" to lower food consumption and to limit transportation. The drop in world prices of sugar and oil and the low exchange rates of the US dollar contributed to this crisis. Finally, in June 1987 Cuba decided to stop payments of capital and interests on its external debt, which by now had reached $6.2 billion.[4]

Between 1989 and 1991 Cuba's situation worsened as the Soviet Union began to fragment. In 1989, in Hungary, its communist party with liberal tendencies came into power and soon proclaimed the Republic of Hungary. In Poland the Solidarity movement won with the support of the Roman Catholic Church and of the former cardinal archbishop of Krakow Karol Wojtyla, His Holiness Pope John Paul II.[7] On November 9, 1989, the Berlin Wall fell, and so had the communist governments of Bulgaria, Romania, and Czechoslovakia by the end of the same year. Germany unified on August 23, 1990, and thus the German Democratic Republic

(GDR) vanished. In 1991 the Baltic States of Estonia, Lithuania, and Latvia declared their independence from the Soviet Union and elected democratic governments. On December 26, 1991, the Soviet Union was formally dissolved 1 day after the resignation of the Soviet Premier Mikhail Gorbachev, author of the *perestroika* (reform) and *glasnost* (liberalization) policies.[8]

By the beginning of 1992 it was evident that Cuba—the only communist country in the American continent—had become almost completely dependent on the Soviet Union and former Socialist Countries.[9] As Cuba's principal export partner, the Soviet Union had received in exchange for these goods almost all of Cuba's sugar, citrus fruits, nickel, and pharmaceuticals. But without the Soviet support, these products could not be produced for export. According to Fogel and Rosental,[9] sugar production relied upon equipment produced in Poland, the GDR, and Czechoslovakia. From the GDR came also the sulfur used for nickel production, the sodium carbonate needed to produce glass, and the alum required for water purification. The GDR also supplied the bulk of Cuba's paints, tires, animal feed and fodder, powdered milk, and frozen poultry meat. Hungary provided buses. From Bulgaria came wheat, cheese and other dairy products, oils, electric motors, and home appliances. Czechoslovakia exported airplanes, communications and computer equipment, shoes, leather, and the malt needed for beer production. Cuba depended upon Albania for other food products and on Romania for heavy equipment and trucks.[9]

As the power of the Soviet Union dwindled, so did Cuba's reserves of currency and food and its ability to obtain credit from the international credit organizations. To make matters worse, the 30-year-long US economic embargo against Cuba was tightened as a result of the Torricelli bill—The Cuban Democracy Act of 1992—which prohibited US companies and even third-country subsidiaries of US companies from trading with Cuba.[10] The US economic embargo became, in effect, a blockade.

Cuba then reached the "zero option"—the peace-time special period, *el período especial en tiempos de paz*.[11] This was originally a war-time plan to survive a potential military blockade to the island, during which severe rationing of food would be implemented, transportation would cease, and nonessential industries would close, as a prelude to the massive migration of people to the countryside and to the mountains in order to survive.

The effects of the *período especial* were palpable everywhere. There were constant blackouts and electric energy outages and an almost total absence of traffic on the highways. In the city of Havana, 2.1 million workers were laid off from work. The crowded city was silent and sad, and nothing moved under the tropical sun. On the wide avenues, heavy black bicycles imported from China, with names such as Forever, Flying Pigeon, and Phoenix, competed for space with horse-drawn carts. On the sidewalks long lines of women and elderly people with empty baskets hanging on their arms waited for their meager food rations of rice, beans, and sugar. With no feed for livestock and no fertilizers and pesticides for farming, and with yokes of oxen tilling the fields instead of tractors, dairy products, poultry, eggs, beef, pork, coffee, tea, chocolate, fruits, vegetables, and cooking oil became unavailable. A stark background of ramshackle houses, piles of rubble, and absence of commerce highlighted Cuba's slow decay. The US dollar again became lord of the island, and black market activity, petty crime, robberies, and prostitution increased significantly.

People lingered on streets, hopeless and listless, which seemed inconsistent with the usual tropical rhythm of life in the Caribbean. However, the Cuban people's faces reflected a quiet resignation and discipline, their stance was proud, and their attitude showed the will to endure—true Spanish *pundonor*,[12] the point of honor that led these people to be stoic through their lot, disregarding the odds and the possibility of defeat.

ENDNOTES

1. Fidel Castro at the Congreso de los Trabajadores de la Agricultura y la Silvicultura, 22 November 1991 (*Havana Cuba Vision Network* FBIS-LAT-91-230, 29 Nov. 1991). Cited by Jean-François Fogel and Bertrand Rosenthal. *Fin de Siglo en La Habana, Los Secretos del Derrumbe de Fidel* (Bogotá, 1994).
2. Agencia Internacional de Noticias (AIN), 18 September 1992; quoted by Fogel and Rosenthal (op. cit.).
3. CMEA = Council on Mutual Economic Assistance; it included Hungary, Czechoslovakia, Poland, Bulgaria, Romania, and the German Democratic Republic.
4. The Economist Intelligence Unit: *Cuba, Country Profile 1992–1993* (London, 1992).
5. According to Fogel & Rosenthal (op. cit.) Soviet military aid to Cuba ranged between $900 and $1500 million dollars a year, to support 145,000 soldiers, 110,000 reservists, 35,000 members of the Air and Naval forces, 1100 tanks, 1400 artillery pieces, and more than 300 Russian MiG and Czech airplanes.
6. In 1970, according to Robert E. Quirk: *Cuba in Africa*, in *Fidel Castro* (New York, 1993), Cuba began sending technical assistance, represented by doctors, teachers, engineers, agricultural technicians, and athletic coaches, to small African nations such as Guinea-Bissau, Equatorial Guinea, São Tomé, Principe and the Seychelles. By 1976, during the Ford administration, about 15,000 Cuban troops were sent into combat in Angola and this number reached 50,000 in 1988. According to Fogel and Rosenthal (op. cit.) in May of 1991 when the last "international volunteers" were repatriated from Angola, Cuban human losses in Africa reached 2289 lives. According to Fogel & Rosenthal

(op. cit.) during 13 years of war in Angola, Cuba participated with an army of upto 50,000 men (in 1988) including 19 regiments of motorized infantry, a small infantry contingent, two anti-aircraft batteries, and one thousand military instructors and advisors.

7. Carl Bernstein and Marco Politi. *His Holiness John Paul II and the hidden history of our time* (New York, 1966).

8. Henry Kissinger. *Diplomacy* (New York, 1994).

9. According to Fogel & Rosenthal (op. cit.) from the old Soviet bloc came Cuba's only supply of crude oil, rice, peas, vegetable oil, condensed milk, butter, packed meat, canned goods, flour, cereal grains, fertilizers, automobiles, tractors, bicycles, TV sets, refrigerators, watches, sewing machines, washing machines, wood, paper, soap, and detergents.

10. The Cuban Democracy Act (CDA) was presented to the U.S. Congress by Congressman Robert G. Torricelli (Democrat, New Jersey) and Senator Bob Graham (Democrat, Florida). www.treasury.gov/resource-center/sanctions/Documents/cda.pdf.

11. *Período especial en tiempos de paz* (Wikipedia, la enciclopedia libre). Cuba's Gross Domestic Product (GDP) decreased 36% during the years 1990–1993.

12. *Pundonor* = James A. Michener, in *Iberia: Spanish Travels and Reflections* (New York, 1968) offers a definition of *pundonor*: "…it has been left to Spain to cultivate not only the world's most austere definition of honor, but also to invent a special word to cover that definition. Of course, Spanish has the word *honor*, which means roughly what it does in French, but also the word *pundonor*, which is a contraction of *punto de honor* [point of honor]."

CHAPTER 3

Ariza Prison
Cienfuegos, Cuba, August 5, 1992

And I, woe to me, imprisoned in my cage, the great battle of men do watch.
José Martí

José Tomás Solano Echegaray,[1] the patient in bed #7, died suddenly and without a diagnosis during the night of August 5, 1992. Dr Laila Gómez,[1] the Chief Resident in charge of the Internal Medicine service at the Provincial Hospital "Aldereguía" in Cienfuegos—a serious young Cuban woman, tall, blond, with almond eyes and a caramel complexion—had done everything in her knowledge to find an explanation for her patient's enigmatic illness.

The patient was a 49-year-old man, an inmate transferred 1 week earlier from the infirmary of the *Reclusorio de Ariza*—a prison facility located in the small town of Ariza about 15 km north of Cienfuegos. The food at the Ariza prison had been reduced in the past 2 months to a salty broth in which small pieces of plantain, cassava, a lost black bean, or even a morsel of fish might be seen occasionally. Prisoner Solano Echegaray had been a fastidious eater because of his ulcer, and in the last week of his life he had only consumed a few sips of soup. Moreover, he had developed a persistent liquid diarrhea, and his fellow prisoners had tried to help him by giving him tepid sugar water until his transfer to the prison's infirmary. His feet and legs had swollen to the point where he could no longer put on his trousers and much less his shoes. He could hardly walk because he was too weak to lift his feet, and the burning pain in the soles of his feet kept him awake at night. After a day in the inmates' infirmary he was sent to the Provincial Hospital. Dr Gómez had followed the usual approach used in the investigation of similar cases of leg edema, searching for the obstruction of veins and lymphatic channels usually seen in cancer patients, or ruling out cardiac problems, low levels of proteins in blood, and kidney failure.

Cuban Blindness. DOI: http://dx.doi.org/10.1016/B978-0-12-804083-6.00003-5

The search had been slow and unrewarding. Power outages in the hospital, lack of laboratory reagents, and limited medications made it difficult to identify a medical problem, which only a few years earlier, when Dr Gómez was a student at the Medical School of the Province of Cienfuegos, would have been done in a single day.

To compensate for the missing laboratory facilities, Dr Gómez had examined her patient slowly and meticulously. She had questioned him thoroughly, persisting in her effort to find a clue that could help her name her patient's malady. His medical history was simple: He was a heavy smoker—like most inmates—and had been operated on for a bleeding duodenal ulcer 27 years earlier. In a soft sibilant voice that revealed his educated Castilian Spanish accent, he correctly informed her that he had "survived a gastrojejunostomy or Billroth II operation." Dr Gómez remembered that patients missing a part of the stomach develop problems with absorption of cobalamin, which she planned to correct by administering vitamin B_{12} injections and improving his diet.

The prisoner was reluctant to tell the doctor the reason for his long prison sentence. "Counterrevolutionary activities," he mumbled. He looked much older than his years, and his eloquent eyes, straight nose, and the few sparse threads of white hair on his head gave him the solemn look of a Rabbi. He was thin, pale, and had the toothless grin of a sexagenarian. Four or five remaining teeth yellowed by tobacco and eroded by cavities were loose in their sockets, and his gums bled. His blood pressure was low, his pulse was fast, and he was short of breath after walking just a few steps. His lungs had the coarse sounds typical in old smokers, and his heartbeats were rapid but regular. His muscles bulged in a large hernia underneath the scar from his past surgery. Finger tapping over the liver and spleen returned a dull normal sound, and the doctor found no suspicious lumps in the flat thin abdomen. The pulses in his groin were strong, and his skin was cool to the touch with no presence of fever. However, the patient was in constant pain, and he stated that the soles of the feet hurt as if the skin had been peeled away from them.

Prisoner Solano Echegaray's feet and legs had swollen to twice the normal size, and the skin over them looked taut and shiny. Gentle pressure on the skin produced a deep dimple that remained for many minutes. His scrotum was swollen and inflated like a balloon. Dr Gómez thought of the grotesque pictures in textbooks on tropical diseases, especially elephantiasis produced by the worms of filariasis in African patients.

But José Tomás Solano Echegaray had never been to Africa. During the 1970s, when Cuba provided military and technical assistance to

Angola and other African nations, he was already in prison.[2] Nonetheless Dr Gómez took a midnight sample of blood looking for the microscopic parasites of filariasis that clog the lymphatic channels and go out for a midnight stroll in the blood of infected patients. The test was negative for microfilarias.

A few other simple tests—blood count, sedimentation rate, glucose, and blood urea nitrogen—were normal. An electrocardiogram showed signs of cardiac overload. Chest radiography could not be performed, since the films were strictly rationed for emergency trauma and surgical cases. Examining the urine proved to be a major challenge because the reagent-strips that measure the presence of protein, blood, glucose, and a number of other elements in urine were produced by the US firms and could not be imported by Cuba.

Dr Gómez was almost certain that her patient in bed #7 had nephrotic syndrome. She had seen children and adults bloat like puffer fish when the sieves of their kidneys slackened, letting proteins escape into urine. Since the central laboratory of the hospital was always overwhelmed with work, she decided to examine the translucent, amber-colored sample of urine in the laboratory of the renal unit on the second floor.

She opened the specimen bottle near the blue flame of a Bunsen burner, heated the end of a thin metal loop to sterilize it, and then dipped it into the sample of urine, which caused a brief sizzling sound. Then she placed a drop of urine on a glass slide and covered it. She stared through the binocular microscope expecting to find blood cells and the telltale noodle-like cylinders formed by proteins. But she found only a clean surface and a few normal epithelial cells. She was surprised by what she found, and then she poured a urine sample into a test tube and centrifuged it for 5 min, discarded the top portion, added two drops of a violet stain to the bottom sediment, and examined it again under the microscope using phase contrast. Again, she found nothing.

The chief laboratory technician confirmed her negative findings, and Dr Gómez then asked her to measure the protein in the urine sample by using the cumbersome chemical method that involves trichloroacetic acid and the biuret reaction. By that afternoon, the results were conclusive: no protein in the urine. The disease of her inmate was not due to a renal problem.

Dr Gómez gave the patient B_{12} injections and paregoric for pain, and to treat the edema of the legs, she began giving him small doses of a diuretic, which caused him to urinate frequently. In 2 days his swelling had come down, and Solano Echegaray was walking more easily. Perhaps because he felt better—or maybe because he had the premonition that death was close—he gave Dr Gómez a succinct description of how he became a prisoner. He was a college student on January 1, 1959, when Fidel Castro and his *barbudos*, his bearded troops, arrived triumphantly in Havana. In April of that year they appeared at the United Nations general assembly in New York.[3] Castro was struggling to make Cuba independent from the United States, only to fall under the Soviet influence.[4] In February 1960, Soviet Deputy Premier Mikoyan visited Havana to sign the USSR–Cuba pact; José Tomás Solano Echegaray was part of a group of Catholic students who had protested against the Soviets' visit. He was arrested by the police and spent several days in jail. Shortly after he returned to the recently reopened University of Havana, a fellow student asked him if he would be willing to participate in the efforts to liberate the country from the threat of Communism with the help of the United States. He became a CIA informant and was again arrested in April 1961 during the roundup of suspected collaborators preceding the Bay of Pigs invasion. At 18 years of age, José Tomás Solano Echegaray was condemned to 35 years in prison for counterrevolutionary activities which were considered treason against the Cuban homeland.

Dr Gómez remembered that as a schoolgirl—dressed in a red skirt, white blouse, and the blue bandanna of the pioneers' uniform—she had learned the basic facts of the failed Bay of Pigs invasion, known in Cuba as *La Batalla de Playa Girón*, the battle of Girón Beach.[5]

Excited and exhausted, Prisoner Solano Echegaray stopped talking and went into a long period of silence. Dr Gómez got up to leave and thanked him for confiding in her. He said in somber conclusion: "It is the story of a lost life; and I still have 4 more years to serve…."

That same night, Solano's heart failed. The intern on call treated the patient with intravenous digitalis, potassium, and diuretics. Despite these measures, the patient drowned in the pink froth that filled his lungs. The next morning, Dr Gómez arrived at the medical ward and was jolted by grief when she saw, in the ambiguous light of dawn, the rolled mattress of bed #7.

Prisoner Solano Echegaray was not the only one to arrive from the Ariza prison. By the end of August 1992, a total of 22 inmates had been admitted to the Aldereguía Provincial Hospital, all of them suffering from swelling and pain in their feet and legs, complaining of difficulty walking, weakness, and fatigue. Apparently, their medical problems had been precipitated by an outbreak of intestinal illness at the prison. All the inmates had lost weight as a result of the poor prison diet and the constant diarrhea. Dr Alfredo Espinoza, professor of Internal Medicine, and his colleagues Drs Apolinaire and Alvarez reached a diagnosis of beriberi.[6] They promptly instituted the appropriate treatment with vitamin B_1 (or thiamine) and rapidly obtained a cure in all of the remaining patients. Also, at their request, the meager diet of the inmates at the Ariza prison was supplemented with a multivitamin pill, effectively preventing the occurrence of new cases.

Professor Israel Borrajero, an experienced pathologist and director of the Pathology Reference Laboratory at the Hermanos Ameijeiras Hospital in Havana, concluded that inmate José Tomás Solano Echegaray had died of beriberi.[7] Professor Borrajero found extensive heart lesions, fragmentation of the myocardial fibers, and presence of vacuoles, as well as severe acute pulmonary edema. Also, the peripheral nerves showed axonal degeneration, with predominant loss of large myelinated fibers typical of the neuropathy of beriberi.

Beriberi is an ancient disease caused by lack of thiamine and a sugar-rich diet that has been eradicated from most countries in the world—an almost forgotten disease of poor rice-eaters in the tropics, which targets and destroys heart muscles, nerve fibers, and the brain. Beriberi is usually the first disease to occur during conditions of famine because the human body has minimal deposits of thiamine in storage. The resurgence of beriberi in Cuba in July and August 1992 among the population of inmates described here illustrates that fact that some tropical diseases have not been eliminated but lie dormant waiting for the right conditions to return.

ENDNOTES

1. The names are fictitious but the clinical history is real and based on the autopsy study of a fatal case of a prisoner from the Ariza Prison. The diagnosis of beriberi was confirmed by Dr Israel Borrajero at the National Reference Center for Anatomic Pathology at the "Hermanos Ameijeiras" Hospital in Havana.
2. According to Robert E. Quirk, *Cuba in Africa*. In, *Fidel Castro* (New York, 1993), Cuban troops were first sent to Angola in 1976.
3. According to Tad Szulc. *Fidel, a critical portrait* (New York, 1986) and Robert E. Quirk (op. cit.) the memorable visit to the United Nations took place from April 15 to 30, 1959.
4. General Eisenhower and his Vice President Richard Nixon—then in the last months of their administration—decided that the U.S. Central Intelligence Agency (CIA) should set the stage for a change of government in Cuba, based on the success story of the covert actions of the CIA in Guatemala that led to the fall of President Arbenz. During the first years after the triumph of the Revolution in Cuba sugar prices were low, there was unrest, unemployment, nationalization of property, and shortages of food, soap, deodorants, and toiletries. Physicians and other professionals were emigrating to the

United States in large numbers. The Soviets came to the rescue and offered to buy one million tons of sugar annually for the next 4 years, along with technical assistance, bettering the annual American quota of sugar purchases.

5. Bay of Pigs is known in Cuba as the battle of Girón Beach because the invading forces were entering the Bay of Pigs while the Cubans were defending Girón Beach. This failed military invasion of Cuba was undertaken by the CIA-sponsored paramilitary group Brigade 2506 on April 17, 1961 (Wikipedia). On March 17, 1960 President Eisenhower approved a CIA covert action plan against Castro (code name Operation Pluto) following the blueprints of the Guatemalan Arbenz operation; this eventually led to the Bay of Pigs invasion of April 17, 1961, during the early days of the Kennedy administration. See: Robert E. Quirk. *The Bay of Pigs* (op. cit.); Trumbull Higgins. *The perfect failure: Kennedy, Eisenhower, and the CIA at the Bay of Pigs* (New York, 1987); Thomas G. Paterson. *Failing the tests: The United States and Cuba in the Castro era;* in, *Contesting Castro* (New York, 1994); Haynes Johnson. *The Bay of Pigs: The leaders' story of Brigade 2506* (New York, 1964); Peter Wyden. *Bay of Pigs: The untold story* (New York, 1979). According to Wyden (op. cit.), initially 430 men were trained at a camp in Guatemala (Base Trax), 60 went to receive guerrilla training in Panama and the rest formed the 2506 assault brigade. Eventually more than 1400 recruits were trained (Johnson, op. cit.). The brigade's air force was trained by American pilots recruited by the CIA from the Alabama Air National Guard (Wyden, op. cit.; Paterson, op. cit.). The brigade's frogmen, the first to hit the beaches, were trained by the CIA on Vieques Island, near Puerto Rico. According to Szulc (op. cit. p. 613), 1189 men were taken as prisoners after the defeat of the invasion forces at the Bay of Pigs, and 107 lost their lives, the Cubans lost 161 dead in the battle.

6. Espinosa A, Apolinaire JJ, Alvarez O. Algunas consideraciones sobre un brote de neuropatía epidémica en el Reclusorio de Ariza, Provincia de Cienfuegos [Some considerations on an outbreak of epidemic neuropathy in Ariza Prison, Province of Cienfuegos]. *Revista Finlay (Cienfuegos)* 26 de Julio 1993:2–3; Muci Mendoza R. Neuropatía óptica cubana. Parte I. Relato de una vívida experiencia personal [Cuban Optic Neuropathy. Part I. Narrative of a vivid personal experience] *Gaceta Médica de Caracas* 2001;109(2):270–275; Muci Mendoza R. Neuropatía óptica epidémica cubana. Parte II Aspectos neuro-oftalmológicos, neurológicos, nutricionales e históricos [Cuban epidemic optic neuropathy. Part II. Neuro-ophthalmological, neurological, nutritional and historical aspects]. *Gaceta Médica de Caracas* 2002;110(2):188–193.

7. According to Edward B. Vedder (*Beriberi*, New York, 1913), beriberi is an ancient disease described in the *Neiching*, one of the oldest Chinese medical treatises, dating back to 2697 BC. Also in China, the *Han Yu* (800 AD) noted that the disease affected mainly rice-eating people in the south of China but not the wheat-eating populations in the north. In Japan, the *Senkinho* (640 AD) describes beriberi heart disease under the names "kakke" and "shoshin," two terms that persist until today. The Greek geographer Strabo described beriberi among Roman legionnaires invading Arabia in 24 BC.

The first modern description of beriberi dates from 1642 when Jacobus Bontius, a Dutch physician working in Batavia (Indonesia), published the first Tropical Medicine book, De Medicina Indorum (Bontius J. De Medicina Indorum. Lib. IV, Part 3, Caput 1. De paralyseos quadam specie, quam Indigenæ beriberii vocant. Lugduni Batavorum: Apud Franciscum Hackium, 1642). Bontius provided in this book the first coherent description of beriberi, under the heading: "De paralyseos quadam specie, quam

Indigenæ beriberii vocant" (Concerning a type of paralysis that the natives call beriberi). The word *beriberi* originated from Ceylon (Sri Lanka) where the Singhalese word *beri* means weak, and it is repeated to emphasize the extreme weakness of the legs in those affected by the neuropathy of beriberi.

From the seventeenth to the twentieth centuries beriberi was the most important public health problem in the Orient and the major cause of morbidity and mortality among populations depending on polished rice for their staple diet—China, Japan, Indonesia, the Philippines, and India—but it also occurred in Africa and tropical South America. Beriberi was common among children of mothers affected by beriberi. Traditionally, poorly nourished prisoners were preferential victims of beriberi (Robert C. Williams. *Toward the Conquest of Beriberi*. Cambridge, MA, 1961).

Beriberi was also a common occurrence among the military (Vedder, op. cit.). For instance, between 1878 and 1883, beriberi affected between 23% and 40% of the total Japanese Navy force of over 5000 men and caused more deaths than enemy casualties. Dr Kanehiro Takaki (1849–1920), a Japanese Navy physician trained in England, demonstrated that beriberi was due to a nutritional factor. To test his dietary theory of beriberi, Dr Takaki conducted one of the first controlled clinical trials in the history of medicine, where he compared the crew of the training ship Ryujo that received the usual Japanese Navy white-rice diet, with the crew of the training ship Tsukuba that was given an enriched diet with more proteins, condensed milk, and meat. The Ryujo left Japan in December 1882 and the Tsukuba in February 1883; both training ships sailed to New Zealand to continue along the coast of South America from Valparaiso to Lima, and then across the Pacific to Honolulu, and back to Japan in voyages lasting some 9 months. The results were outstanding: Of the 276 crewmen of Ryujo, all of whom were eating the white-rice diet, 169 (61%) contracted beriberi, and 25 died (mortality rate = 15%). In contrast, only 14 (5%) of the crew of Tsukuba who ate Takaki's enriched diet contracted beriberi, and none died. Admiral Takaki eventually eliminated beriberi from the Japanese Navy. However, his recommendations were not followed by the Army, and in 1905, during the Russo-Japanese War, the Japanese Army had 200,000 cases of beriberi (Takaki K. Kakke or Japanese beriberi. *Lancet* 1887;2:189; Takaki K. Three lectures on the preservation of health amongst the personnel of the Japanese Navy and Army. *Lancet* 1906;1: 1369; 1451; 1520).

Beriberi was endemic in the Bataan region of the Philippines at the onset of World War II, and the poor nutrition of the troops defending Bataan and Corregidor probably resulted in beriberi; this condition may have contributed to the 21,600 deaths of Filipino and American troops captured by the Japanese during the Death March of Bataan.

The reason for the persistence of beriberi in the Orient was the milling to produce polished white rice that removes the germ (endosperm and embryo) off the grain where thiamine is stored. In 1897, Christiaan Eijkman (1858–1930) and Gerrit Grijns (1865–1944), working in the Dutch East Indies (Indonesia), discovered that polyneuritis gallinarum could be induced in chickens and pigeons by a diet restricted to polished rice. Later, polyneuritis gallinarum was found to be identical to human beriberi. Eijkman and F. Hopkins were awarded the 1929 Nobel Prize in Medicine for their discoveries. In 1910, Capt. Edward B. Vedder USAMC, stationed in Manila, the Philippines, was the first to use an antineuritic rice extract—containing thiamine—to successfully treat children with beriberi. In 1911, Casimir Funk isolated a pyrimidine-related concentrate (mostly niacin) from rice polishings that cured polyneuritis in pigeons. He named the concentrate "vitamine" because it appeared to be vital to life and because it was probably an amine. Also in the Philippines, Robert Williams, a chemist, worked with Vedder in Manila to characterize and isolate the active substance in the antineuritic extract from rice polishings. Williams returned to the United States in 1915 and pursued his quest for many years, unsuccessfully. In 1927, the antineuritic vitamin was isolated by Jansen and Donat. However, in 1933, Williams and coworkers obtained the crystalline form and in 1936 synthesized thiamine, the anti-beriberi vitamin and the first of the B-group vitamins. Williams obtained US Patent Number 2,049,988 for his method to produce industrial quantities of thiamine at low cost. In 1958, FAO/WHO recommended partial milling of rice and worldwide enrichment of rice, white flour, and bread with synthetic thiamine as the main steps to eradicate beriberi (Hardy A: *Beriberi, vitamin B1 and world food policy*, 1915–1970. *Medical History* 1995;39:61–77; Williams, op. cit.)

CHAPTER 4

The Strange Epidemic
Geneva, May 4, 1993

*Nothing strikes the mind with greater wonder than
those diseases which we call epidemic.*
Thomas Sydenham (1679)

Like raindrops on a tropical island, cases of blindness began to appear here
and there under the sunny sky. The blindness occurred sporadically, and no
one could have anticipated the fury of the storm about to ravage Cuba.
In November 1991 when he became ill, José Tomás Polo was one of the
first rare cases of blindness in his town, but during the next several weeks
he heard of similar cases all over the region. Early in February 1992 his
daughter told him that three other people of his age, Manuel Coro, Benito
Fuentes, and Pepe Casas, were going blind.

By May 1992 there were rumors that almost 100 cases had been diagnosed in Pinar. José Tomás knew many of the tobacco *guajiros* affected. The
woman in charge of the Committee for the Defense of the Revolution in
his hometown had visited him often to remind him to keep his follow-
up appointments for treatment at the Regional Hospital. She told him
that there were other blind men in many of the smallest towns: 18 cases
in Consolación del Sur, 15 in San Juan y Martínez, 5 in Güane, and 3 in
the tiny town of San Luis. There were also rumors that cases of blindness had appeared in Havana and in other large cities of Cuba. Nobody
knew what the cause of the problem was, and the random selection of
those afflicted with blindness was even more puzzling. The doctors at the
hospital in Pinar attempted several treatments; first with antiinflammatory
steroids and then with injections of vitamins. The latter had improved José
Tomás Polo' eyesight a little and prevented the condition from getting
worse, but in practical terms his blindness remained unchanged.

There were many rumors in the island about the nature of this strange
illness. Some believed the enemy had introduced it, some that it was the
result of atomic radiation waste from the Soviets, and some that it was

Cuban Blindness. DOI: http://dx.doi.org/10.1016/B978-0-12-804083-6.00004-7

caused by the ground meat mixed with the soy protein from China, which had replaced the traditional Cuban *picadillo*.[1]

Date: 1/15/1993 Bethesda: I received this morning in my office a telephone call from the Cuban Trade Commissioner in Canada. The person at the other end spoke poor English with a strong Caribbean accent. He identified himself as Carlos Barros from the Cuban Consulate in Montreal. I replied in Spanish, and he began by stating that he was not a physician and that a doctor at the Montreal Neurological Institute had recommended my name to him.

"Dr Román," he said, "I need to mention to you that this is confidential information. Cuba is now fighting an epidemic of blindness. The doctors in Cuba think that the disease is optic neuritis."

Astonished, I asked him, "How many cases have occurred?"

"Several hundred…," Mr Barros replied vaguely and then continued, "Because of the problems with the US embargo, it is very difficult for our doctors in the island to find scientific information. Our consular services have been given the urgent task of obtaining as much material as possible about epidemic blindness and optic neuritis. Could you help us by providing some articles on this problem? Send the scientific papers to me, and I will make sure that they reach the specialists in Cuba." I expressed my sympathy to the Cuban diplomat, offered my collaboration, and promised to comply with his request.

"Several hundred cases of blindness in Cuba…," I said to myself. "What could be the cause?" I mentally performed a rapid differential diagnosis and placed intoxication as the first possibility, specifically, poisoning with alcoholic beverages adulterated with methanol.[2] I had seen patients in the Caribbean who woke up literally blinded by the hangover after a night of drinking and rumba. Only the few friends sharing a wicked adulterated bottle become blind. These cases typically occur in small clusters, and it was difficult to imagine hundreds of people blinded by methanol.

Optic neuritis is an inflammation of the optic nerves associated often with multiple sclerosis, but it sounded implausible in this case. Multiple sclerosis is a disease of Nordic Scandinavian countries, truly rare in the tropics, where abundant sunlight and vitamin D appear to protect people against autoimmune diseases.[3] Therefore optic neuritis and multiple sclerosis were highly improbable in Cuba, and I knew of no reports of epidemic optic neuritis in that region.

Finally, there were the nutritional diseases that are common in the tropics. Poor countries, usually lumped together as the Third World, are located preferentially in tropical and subtropical regions and are inhabited by one-third of the population of the world but consume a mere 6% of the global food production, making them prone to diseases of hunger. I had conducted extensive research on the neurologic epidemics of the tropics—including blindness—and had published in 1986 a Spanish-language paper on this topic in the *PAHO Bulletin*.[4] I retrieved a stack of reprints of this paper and also included the comprehensive English version of my review in the envelope addressed to Mr Barros in Montreal, along with copies of a case-report on methanol from my files.[2]

I wondered if this epidemic of blindness could be a nutritional problem. In the past, hundreds of cases had been reported under special conditions of nutritional deprivation, but the exact magnitude of the neurology of malnutrition in the tropics was unknown, simply because there are no neurologists and no optic nerve specialists in the poor countries where these problems occur. It was exactly for these reasons that I had coined the term "hidden endemias" for these problems of tropical neurology.

Cuba has been, for years, a hidden territory, and I have no idea of the scientific level of their specialists in neurology, ophthalmology, and epidemiology, but this problem of blindness presents a challenge worthy of intensive study.

Official Announcement: Optic Neuritis in Cuba

The first public information on the epidemic of optic neuritis in Cuba appeared on Saturday, April 3, 1993. A communiqué from the Cuban Ministry of Public Health, MINSAP, was published in *Granma*, the Communist Party newspaper, and was reprinted the next day in *Juventud Rebelde*—the only Sunday newspaper in Cuba.[5] It was printed under the headline "Information on Optic Neuritis" and was promptly made available to the international press by a Reuters cable that summarized the lengthy communiqué as follows:[6]

Cuba Hit by Epidemic Linked to Vitamin Deficiency
Dateline: HAVANA
Priority: RUSH
North American News Report
Byline: Pascal Fletcher

Cuba said on Saturday it has been hit by an epidemic causing gradual loss of sight with lack of vitamins and poor nourishment among the main aggravating factors. "An eye illness with the characteristics of an epidemic has appeared recently in our country," the Health Ministry said in a statement published by the Communist Party newspaper Granma.

Communist-ruled Cuba is currently in the midst of its worst-ever economic crisis marked by acute shortages of food, medicines, fuel, and basic consumer goods.

In announcing the epidemic, the Health Ministry also strongly attacked the United States, saying a long-standing US economic embargo against Cuba "contributed cruelly to the spread of this illness" by making it more difficult to obtain medicines and food. The ministry identified the illness as "optic neuritis," saying it gradually and progressively affected sight in both eyes. It had been detected in all of Cuba's 14 provinces after appearing in the western province of Pinar del Río last year. The current outbreak in Cuba was "unusual in its extent and acuteness," the ministry added, although it did not give the total number of cases.

In addition to saying the United States and its embargo were "the main ally of this epidemic and its consequences," the ministry said specialists were still investigating the possibility that the illness might have been deliberately introduced onto the island.

With the *Granma* and *Juventud Rebelde* reports, the Cuban population first found the explanation for the strange symptoms that were becoming so frequent on the island. Fatigue, weight loss, lack of drive and energy, palpitations, mouth sores, muscle aches and pains, numbness in the hands, and a sensation of burning in the feet that prevented sleep at night—all these had inexplicably surfaced. Above all, a wave of irritability and malaise had spread throughout the island replacing the normally festive Cuban temperament. For the first time in many years there were open signs of dissatisfaction on the island. Signs reading "¡Abajo Fidel!" and "¡Fidel traidor!"—"Down with Fidel!" and "Fidel is a traitor!" began appearing alongside the elaborate murals of pro-Revolution slogans. The population was tired and hungry.

Rapidly, the government began limiting the manifestations of discontent and emphasized that a toxic factor could be causing the epidemic.

A Reuters report on April 19, 1993 stated:[7]

Health authorities, without offering precise details, say several thousand people across Cuba are suffering from "optic neuritis," a disease … which causes progressive loss of sight in both eyes. A senior doctor at one of Havana's hospitals treating the disease said on Monday that a team of specialists was engaged in an intensive investigation of the illness to pinpoint its exact causes, including the possibility that it might have been deliberately introduced to the island by Cuba's enemies … Large numbers of anxious people are flocking daily to hospitals to be checked for the illness, which has coincided with a nationwide economic recession marked by severe shortages of basic foods like fresh meat, milk and eggs…

El período especial, the special period of wartime austerity, began in 1991 with the disappearance from the bodegas of beef, pork, fish, chicken, eggs, milk, butter, cheese, bacon, and olive oil. Traditional bakeries became extinct, and bread, coffee, and chocolate were severely rationed, and tea vanished. An odd-tasting mixture of soybean meal and meat or fish scraps called *picadillo* became the only source of animal protein. For the next years, rice and beans would become the staple diet of every Cuban. Only children under 15 received milk and eggs. Pregnant women and those 65 and older continued to receive oral vitamins and a protein supplement.

In addition to suggesting a foreign source for the epidemic, and per-haps to discourage the idea that bad nutrition was a contributing factor, Cuban authorities said the major risk factors for the development of the neuropathy were alcohol consumption and smoking. Most Cubans failed to accept the official version: "Many Cubans have reacted with anger and incredulity to the official position," Reuters reported.[7] "'People have been drinking and smoking in Cuba for years and they've never suffered any-thing like this,' one Havana resident said. Most ordinary Cubans have no hesitation in linking the illness to the food situation. They say the daily intake of calories and vitamins has fallen dramatically over the last two years as the food shortages have gotten worse." According to Reuters:[7]

> Despite their reluctance to link the epidemic to poor levels of nourishment, author-ities are treating affected patients with intensive doses of vitamins and distribut-ing vitamin supplements to prevent the spread of the disease. Diplomats said that authorities were clearly embarrassed by the illness. It was the first indication that the health of the Cubans was being affected by an economic crisis triggered by the collapse of Cuba's trade and aid ties with Eastern Europe and the former Soviet Union.

The time had arrived to expose the Cuban epidemic before all the nations of the world.

PALACE OF THE NATIONS, GENEVA, SWITZERLAND
46th World Health Assembly, May 4, 1993

Dr Jorge Antelo Pérez, the energetic Cuban Deputy Minister of Health for International Affairs, appeared concerned and deeply embarrassed a few days earlier when he requested permission to change the text—already in the hands of the secretariat—of his original address to the plenary meet-ing of the 46th World Health Assembly of the World Health Organization (WHO), the annual gathering of health ministers from around the world.

The reason for the change was the result of conversations held in Cuba between Fidel Castro and Dr Carlyle Guerra de Macedo, the Brazilian

Director General of the Pan American Health Organization (PAHO). Dr Guerra de Macedo was able to convince the reticent Cuban public health authorities that external support was essential to gain control of the epidemic.

Dr Guerra de Macedo's brief visit to Havana between March 16 and March 20, 1993, had left a major and lasting influence on the handling of the epidemic by the Cubans. His trip was originally scheduled to review the possibility of adapting Cuba's successful Family Doctors program, *el Médico de la Familia*, in developing countries.[8] Dr Guerra de Macedo met with Cuban representatives from the Health Delivery Section of MINSAP, as well as with international observers. Unexpectedly, he was whisked off to an interview with Fidel Castro, who informed him of the ongoing epidemic.

Dr Guerra de Macedo expressed his complete surprise. "Comandante," he said, "I had no idea of this problem." He offered the help of PAHO and suggested that the Cubans should inform the WHO at the forthcoming 46th World Health Assembly.

On Tuesday, May 4, 1993, during the fourth Plenary Meeting of the WHO's 46th World Health Assembly, Dr Antelo officially notified the global health community of the occurrence of the epidemic in Cuba and for the first time provided figures concerning its magnitude.[9] In a calm voice, but perspiring profusely and repeatedly wiping his ample forehead in a nervous movement, Dr Antelo told his fellow ministers of health that a total of 25,959 patients had been treated.

A sound of surprise rose from the audience in response to this number. This was a major epidemic. Limiting his presentation to the strictly allotted time, making no comments regarding the possible cause of the epidemic, Dr Antelo sat down with visible relief.

The news of the epidemic shook the global health community. Cuba had been enjoying a well-deserved distinguished position in the WHO as a nation with major accomplishments in the field of public health.[10] In its first 35 years, the Cuban revolution had dedicated major efforts to health and education and had obtained achievements that no other Third World nation could match. The Cuban health system was universal and free. Infant mortality rates were among the lowest in the world on par with industrialized nations. Polio and most other diseases preventable by vaccination had been eradicated long ago. Protein calorie malnutrition had been eliminated since the 1970s. Life expectancy was 75 years, the highest in Latin America. The population of citizens older than 65 years had doubled in 20 years. There was one doctor per 303 Cubans. The program known as *Médico de la Familia* had brought preventive health and nursing services to every community. Only 2 years earlier, in 1990, an epidemic of cholera had spread from Peru across South America, Central America, and the Caribbean. One million people were stricken, and 9000 died, but not a single case had been reported in Cuba, attesting to the solid conditions of environmental sanitation and clean water supply existing there.

According to Dr Antelo, Cuban health officials knew of almost 20,000 cases of the epidemic of blindness—"19,820 cases of predominantly optic forms." Also, Dr Antelo presented another disturbing piece of information: The disease occurred in several different clinical forms, and in addition to the loss of eyesight, it also appeared to cause damage to the nerves and the spinal cord. He concluded: "There were 5547 predominantly peripheral neuropathy forms and 301 other clinical forms."[9]

In Havana, *Granma* published the text of Dr Antelo's speech in its entirety under the heading: "Our government has acted decisively and energetically in fighting this epidemic."[11] In Geneva, Dr Antelo and the members of the International Relations Office of the Cuban MINSAP lobbied intensively before the international delegations to obtain all the needed help, extending an open invitation to experts from the Americas, Europe, and the rest of the world to come to the island of Cuba to help solve the problem of this mysterious epidemic.

The country seemed to be on a streak of bad luck. In addition to having lost its economic trading partners in the Socialist countries, Cuba was in the midst of another and probably the worst economic crisis ever. It had no hard currency, fuel oil reserves were almost exhausted, and foodstuffs were at a dangerously low level. In addition, in March 1993, Cuba suffered extensive damage to harvests and plantations when the "storm of the century" swept across the Caribbean.[12] The storm destroyed more than 20,000 homes on the island, resulting in losses exceeding $1 billion, according to the United Nations.[13] And, to top it all, the Torricelli bill, which tightened the US economic embargo against Cuba, was making things much worse. Given the restrictions of maritime trade of the Cuban Democracy Act of 1992, the Cuban Government considered that the limitations imposed by this tightening of the embargo amounted to a US blockade.

The Cuban delegation met over the next days with a number of European delegations to provide additional information about the epidemic of optic neuropathy and to obtain research support and food relief. The Cubans also met with the United States delegation headed by Dr Antonia

Novello, Surgeon General of the United States Public Health Service (USPHS)—soon to depart with the arrival of the new Clinton administration.[14] Born in Puerto Rico, Surgeon General Novello spoke with a Caribbean Spanish accent that was almost identical to that of the Cubans'. She showed genuine interest in their problems and quickly assured the Cubans that the US government would offer help during this health emergency. The International Red Cross also promised relief and representatives from the European Community—in particular Spain, England, and Italy—announced their intention to send scientific missions to the island and to provide basic materials for production of vitamin supplements.

At the WHO headquarters, the task of answering Cuba's request for help fell on Dr Bjørn Thylefors, Director of the WHO Blindness Prevention Program. The tall and easy-going Scandinavian physician had many years of experience fighting the common causes of blindness in the tropical world: Trachoma, onchocercosis, and nyctalopia (night blindness caused by lack of vitamin A), as well as cataracts, easily treatable but still a cause of permanent blindness in many regions of the world.

In Thylefors' experience, an epidemic of blindness of this magnitude was unprecedented. At a meeting in Geneva, Dr Thylefors, along with PAHO's Deputy Director, Dr Robert Knaus from the United States, and Cuba's Antelo, decided to coordinate the actions for the study and control of the epidemic through PAHO, the WHO's regional office for the Americas based in Washington. The initial step would be to assemble an international team of experts in epidemiology, ophthalmology, neurology, and nutrition—a "Mission to Cuba"—to be formed by Spanish-speaking scientists able to retrieve first-hand information from the epidemic's epicenter. The mission's most urgent task would be to take a closer look at the strange epidemic and give it a name and a diagnosis based on an independent, learned opinion.

ENDNOTES

1. *Picadillo* = traditional dish in Spain, Latin American countries, and the Philippines, made with ground beef, tomatoes, and other ingredients. Cuban *picadillo* includes peppers, onions, garlic, oregano, cumin, tomato sauce, stock, olives, raisins, potatoes, and capers and is usually sautéed in olive oil and white wine (Wikipedia).
2. Methanol or wood alcohol is inexpensive, lacks a particular odor or taste, and is used to adulterate expensive beverages, particularly high-ethanol content drinks such as whisky or vodka. Methanol is also found in homemade alcoholic drinks and in bootleg rum. Intoxication may cause permanent blindness, brain lesions, and death. See, Pendlebury WW, Román GC, Muñoz-Garcia D. Methanol poisoning. *Neurology and Neurosurgery Update Series* 1986; 6 (36): 1–8.

3. Sundström P, Salzer J. Vitamin D and multiple sclerosis—from epidemiology to prevention. *Acta Neurologica Scandinavica* 2015; 132 (Suppl. 199): 56–61.
4. Román GC. Mielopatías y mieloneuropatías tropicales [Tropical myelopathies and myeloneuropathies]. *Boletín de la Oficina Sanitaria Panamericana* PAHO 1986; 101: 452–464.
5. Ministerio de Salud Pública. Información sobre la neuritis óptica, *Juventud Rebelde*, Dominical (April 4, 1993).
6. Fletcher, Pascal. Cuba Hit by Epidemic Linked to Vitamin Deficiency (Continued)

   Cuba has in the past accused the US intelligence services of introducing a fatal epidemic of dengue fever to the island. Unconfirmed reports have been circulating on the island for weeks about an increasing number of illnesses caused by vitamin deficiencies and poor nourishment. These reports had said there were several hundred cases in Havana alone.

   Foreign diplomats in Havana said it was the first public indication that the health of the Cuban population was being affected by the current economic crisis, which was triggered by the collapse of Cuba's past trade and aid ties with Eastern Europe and the former Soviet Union. The Health Ministry statement said the current epidemic mostly affected the adult male population and appeared to have more than one cause. It listed the main "risk factors" directly associated with the appearance of the illness as vitamin deficiency, especially of vitamin B complex, and drinking alcohol and smoking. Insufficient food intake also raised the risks, the ministry added. "Our country has a strong health system which is prepared to face the current situation," the ministry statement said. Cuba has prided itself on its free public health system, which has been praised in the past by international specialists as one of the best in the Third World. Improving public health was a priority policy of Fidel Castro's revolutionary government, which took power in 1959 after a guerrilla uprising that toppled right-wing dictator Fulgencio Batista.

7. Fletcher, Pascal. Cuba Tries to Play Down Food Factor in Eye Epidemic

   Dateline: HAVANA, Priority: RUSH, Reuters World Service, Date: April 19, 1993, 18:22 E.T., Byline: Pascal Fletcher, Words: 00445.

   Cuba's communist authorities, working to check a nationwide epidemic affecting eyesight, are also fighting a widespread conviction among the island's people that lack of adequate food is the sole cause of the illness. Health authorities, without offering precise details, say several thousand people across Cuba are suffering from "optic neuritis," a disease related to vitamin deficiency, which causes progressive loss of sight in both eyes. Large numbers of anxious people are flocking daily to hospitals to be checked for the illness, which has coincided with a nationwide economic recession marked by severe shortages of basic foods like fresh meat, milk and eggs. A senior doctor at one of Havana's hospitals treating the disease said on Monday that a team of specialists was engaged in an intensive investigation of the illness to pinpoint its exact causes, including the possibility that it might have been deliberately introduced to the island by Cuba's enemies. "It's a worrying problem and there are some questions to be answered," said the doctor, who asked not to be identified. Senior Health Ministry officials have declined to give on-the-record interviews about the illness. Publicly announcing the epidemic for the first time last April 3, the Health Ministry listed a deficiency in vitamins, especially vitamin B complex, as one of the three main "risk factors," along with drinking and smoking. Since then the tightly controlled state media have placed extra emphasis on the smoking and drinking factors. The authorities have clearly held back from publicly relating the epidemic directly to food shortages. Many Cubans have reacted with anger and incredulity to the official position. "People have been drinking and smoking in Cuba for years and they've never suffered anything like this," one Havana resident said. Most ordinary Cubans have no hesitation in linking the illness to the food situation. They say the daily intake of calories and vitamins has fallen dramatically over the last 2 years as the food shortages have gotten worse. Despite their reluctance to link the epidemic to poor levels of nourishment, authorities are treating affected patients with intensive doses of vitamins and distributing vitamin supplements to prevent the spread of the disease. Diplomats said that authorities were clearly embarrassed by the illness. It was the first indication that the health of the Cubans was being affected by an economic crisis triggered by the collapse of Cuba's trade and aid ties with Eastern Europe and the former Soviet Union.

8. Julio Teja Pérez, Joachim Van Braunmühl, Jorge Jara Valencia, Miguel A. Márquez: *Cuba's family doctor programme* UNICEF/UNFPA/OPS/OMS/MINSAP (Havana, Cuba, 1992).
9. WHO. Forty-Sixth World Health Assembly, Geneva, 3–14 May 1993 Verbatim Records of Plenary Meetings, Geneva, WHA46/1993/REC/2, A46/VR/4, 1993, pp. 65–67.
10. PAHO. *Health Conditions in the Americas: 1981–1984* (Washington, 1986).
11. *Granma*. Nuestro gobierno ha actuado con decisión y energía para combatir esta epidemia. Texto del discurso pronunciado por el doctor Jorge Antelo, viceministro cubano de Salud Pública, ante el plenario de la 46 Asamblea Mundial de la Salud, que se efectúa en Ginebra, donde informó sobre la neuropatía epidémica (April 13, 1993).
12. Storm of the Century (Wikipedia).
13. *Granma*. Aprobada resolución de ayuda a Cuba por daños de la tormenta del siglo (May 13, 1993).
14. The US delegation included Mr W.D. Broadnax, Deputy Secretary Designate of the Department of Health and Human Services; Dr Audrey F. Manley, Acting Assistant Secretary for Health; Mr H.C. Rodgers, Chargé d'Affairs a.i., Permanent Mission, Geneva, and Dr Antonia Novello, US Surgeon General.

CHAPTER 5

Pan American Health Organization (PAHO)
Washington DC, May 10, 1993

> *Pro Salute Novi Mundi.*
> *For the Health of the New World.*
> **PAHO's motto**

There had already been anxious discussions in Washington about the outbreak of disease in Cuba. American health authorities were quite concerned about the presence of an epidemic of this magnitude only 90 miles from Miami. To some, because of the East-to-West distribution of the epidemic, the outbreak appeared to be a messy leftover from careless disposal of Russian chemical warfare agents or radioactive residues before the Soviets' final return home.[1] Since October 1962, the time of the famous "thirteen days" of the Cuban Missile Crisis, the highest concentration of Russian military installations had been located on the western side of the island opposite the GTMO, the US Guantanamo Bay Naval Base. The names of the Cuban missile sites—San Cristobal, Guanajay, and San Diego de los Baños—were still fresh in the memories of many in the US Department of Defense.[2] The primary concern for the North Americans would be to rule out an accident involving agents of biologic or nerve warfare agents left by the retiring Soviet army.

However, for Dr Frank Young, Rear Admiral of the Public Health Service in charge of the Office of Emergency Preparedness, the possibility of a true viral epidemic appeared even more ominous. The office had been monitoring Cuban cable traffic and radio and television transmissions; the disease was hitting the island violently. The pattern of presentation of the epidemic and the apparent absence of casualties made less likely the accidental release of agents of chemical or biologic warfare. With the example of AIDS spreading all around the globe, Admiral Young was more concerned about a potential novel viral agent, perhaps a new retrovirus or a variant of the Ebola or Marburg filovirus family imported from Africa by some of the thousands of Cuban military personnel who returned to the

Cuban Blindness. DOI: http://dx.doi.org/10.1016/B978-0-12-804083-6.00005-9

island following their military tours of duty in Angola and other African countries.[3] If this was a viral epidemic occurring within a few minutes of air flight time from Miami—with limited, but still constant, daily travel of hundreds of people between Havana and Miami—spreading of the disease to the continental United States was distinctly worrisome.

Soon after returning to Washington from Geneva, Dr Robert Knaus and Pan American Health Organization's (PAHO's) Director General Dr Guerra de Macedo met to study the situation created by the Cuban government's announcement about the huge epidemic. International support to fight the epidemic was mandatory and urgent. But the timing of the epidemic presented a peculiar political problem, which precluded, by law, the possibility of direct financial aid from the United States. Given this scenario, it was possible, the two doctors concluded, that the epidemic could be perceived as a political maneuver invented by the Cuban government to force the lifting of the embargo.

It was mandatory, therefore, to obtain a complete and accurate assessment of the epidemic as soon as possible. Sending a fact-finding team of Spanish-speaking experts would be the most expeditious way to proceed.

PAHO's Division of Health Promotion and Protection (HPP) under the direction of Colombian physician Dr Helena Restrepo was given the responsibility to assemble the team and to steer the Mission to Cuba through the maze of bureaucracy, which would be needed to complete the assignment. Dr Restrepo and Dr Guillermo Llanos, Coordinator of HPP's Health Promotion and Social Communication Program, began the delicate process of selecting potential candidates for the team. It was agreed that Dr Llanos would be the PAHO/WHO official representative in the team traveling to Cuba. Dr Guillermo Llanos, a Colombian-born physician with long experience in public health, was a respected university professor of epidemiology, statistics, and public health at the National University of Colombia in Bogotá and at the Universidad del Valle in Cali, Colombia. While there, he developed a brilliant academic program in cooperation with Tulane University's tropical medicine program. After joining the PAHO, Dr Llanos had worked in numerous public health programs in Peru and Central America. His skills in epidemiology would be very helpful in reviewing and analyzing the Cuban data on the epidemic.

Because blindness and loss of vision appeared to be the principal symptoms of the epidemic neuritis, the first person selected for the team was Dr Juan Carlos Silva, Director for the Americas for WHO's Blindness Prevention Program (BPP). Dr Silva was a young physician, who was the son of a prominent surgeon; he was born in Bogotá, Colombia, where he studied medicine and specialized in ophthalmology at the Universidad del Rosario. He had recently completed a fellowship in eye pathology at the prestigious Wilmer Eye Institute of Johns Hopkins University in Baltimore, Maryland. After returning to his home country, Dr Silva joined WHO's BPP and was directing the Latin American program from the headquarters in Bogotá. The Mission to Cuba would be his first assignment in a major international emergency.

Also selected for the team among the roster of PAHO's ophthalmology consultants was Dr Rafael Muci-Mendoza. Born in Venezuela, Dr Muci-Mendoza was Professor of Internal Medicine at the Central University in Caracas, Venezuela, where he was popular among medical students for the accuracy of his diagnoses and his entertaining lectures. His small frame and serious bespectacled face disguised a sharp sense of humor, but his perfectly trimmed bushy white beard gave him the severe demeanor of a European "savant" transplanted from the 1800s. Dr Muci-Mendoza had trained in the United States and was a leading authority in neuroophthalmology, the complex field of study of the visual manifestations of diseases affecting the nervous system.

In the field of nutrition, the choice was clear: Dr Benjamín Caballero, Director of the Division of Human Nutrition at Johns Hopkins School of Hygiene and Public Health in Baltimore, Maryland.[4] Caballero was one of the best representatives of the intellectual elite of Latinos who have conquered the academic world in the United States with their combination of brains, wit, hard work, and that intangible element of success in research that Pasteur called, "the prepared mind"—the gift that would allow the researcher "to walk where others had walked before and see what nobody before had been able to see."[5]

Dr Caballero grew up in Argentina and graduated in medicine in Buenos Aires, the city of wide avenues, European flair, and the voluptuous tango. He was in his early 40s, of elegant countenance, receding dark hair, and a prematurely gray beard. Despite his many years of work and residence in the United States, he still had the *porteño* accent of his hometown. After 7 years in Boston at the Massachusetts Institute of Technology (MIT—Harvard), he obtained his PhD degree, became Assistant Professor of Pediatrics at Harvard Medical School, and was picked to be Assistant Director of the Clinical Research Center at the MIT. In 1990, he achieved a researcher's golden status when he was given his own laboratories and an independent program.

At the young age of 42, Dr Caballero was appointed Director of the Division of Human Nutrition and of the Center for Human Nutrition, Associate Professor of International Health, Maternal and Child Health at the School of Hygiene and Public Health, and Associate Professor of Pediatrics at the School of Medicine of the Johns Hopkins University in Baltimore.

It was agreed that Dr Caballero—a frequent consultant for the PAHO—would be a crucial member of the Mission to Cuba. The only problem was his hectic agenda. Previous assignments for the PAHO had required a 1-year advance notice. But Llanos caught him at a perfect time. Dr Caballero was preparing a grant request to the National Institutes of Health (NIH) for a multimillion-dollar center, and he was in protected isolation at his office. His secretary put Llanos' call through after hearing the words "epidemic," "Cuba," and "World Health Organization." Dr Caballero immediately agreed to join the Latin American team. The Mission was scheduled to leave for Cuba on Saturday May 15. Dr Caballero looked at the calendar. It was Monday May 10.

For reasons that were then unclear, Cuban Deputy Minister Dr Antelo had requested the presence of a virology expert in the team. But the choice of a virologist dedicated to the study of viruses that are capable of invading the nervous system was difficult. Not many virologists were fully fluent in Spanish, very few were experts in viruses affecting the nervous system, and many of the latter were unavailable. After consultation with several PAHO members involved in the fields of vaccines and viral diseases, Drs Restrepo and Llanos focused on Dr Milford (Peter) Hatch, a virologist formerly at the Center for Diseases Control and Prevention (CDC) in Atlanta. Dr Hatch was willing to join the Mission, but ironically, at that time he was recovering from the most frequent of all viral diseases—the common flu. He assured Dr Llanos that he would travel at a later date to join the team in Cuba. In fact, he never went to Cuba.

ENDNOTES

1. In 1962, the number of Soviet troops on the island had peaked at more than 42,000 but toward the end of 1991 the estimated military presence in Cuba was around 7600 according to *Time* Magazine, *Cuba: So long, amigos* (September 23, 1991); see also, Andres Oppenheimer. *Castro's final hour* (New York, 1992).
2. Robert F. Kennedy. *Thirteen days: A memoir of the Cuban missile crisis* (New York, 1969); Herbert S. Dinerstein. *The making of a missile crisis: October 1962* (Baltimore, 1976); Dino A. Brugioni. *Eyeball to eyeball: The inside story of the Cuban missile crisis* (New York, 1991); Central Intelligence Agency. *The secret Cuban missile crisis documents* (Washington, 1994).
3. Peter Brookesmith. *Biohazard, the hot zone and beyond: Mankind's battle against deadly disease* (New York, 1997); Richard Preston. *The hot zone: The terrifying true story of the origins of the Ebola virus* (New York, 1994); Randy Shilts. *And the band played on: Politics, people,*

and the AIDS epidemic (New York, 1987); Peter Piot. *No time to lose: A life in pursuit of deadly viruses* (New York, 2012); David Quammen. *Spillover: Animal infections and the next human pandemic* (New York, 2012). For the recent Ebola outbreak, see: http://www.who.int/mediacentre/news/ebola/archive/en/.

4. Dr Benjamín Caballero graduated as a physician from the University of Buenos Aires, specialized in pediatrics, and in 1980 was awarded a Fellowship to study at the Institute of Nutrition of Central America and Panama (INCAP) in Guatemala. INCAP was a household name in Latin America, known everywhere in the region for its connection to *Incaparina,* a popular baby food—produced by INCAP but sold in the smallest villages and humblest of *tiendas. Incaparina* provided infants with a rich source of essential amino acids and was arguably the single most effective weapon in the fight against childhood malnutrition in Latin America.

    Dr Caballero's career took him from city life in sophisticated Buenos Aires to the Maya-Quiché Amerindian villages on the slopes of the snow-capped volcanoes in the highlands of Guatemala. This experience taught him the true life-and-death importance of nutrition in human life. His career was fast moving. In just two frantic years, he completed with honors his Master of Science degree in Human Nutrition, courted and married a young American journalist on assignment to cover the bloody civil wars ravaging Central America, and obtained a Fellowship to work toward a PhD degree at MIT, the Massachusetts Institute of Technology and Harvard University in Boston. There, Dr Caballero began studying obesity, a field at the other extreme of his previous Latin American experience, and probably the most serious nutritional problem of the North American society of today. Obesity now affects over 50% of adult Americans. See, Caballero B. Insulin resistance and amino acid metabolism in obesity. *Ann N Y Acad Sci.* 1987;499:84–93; Caballero B. Brain serotonin and carbohydrate craving in obesity. *Int J Obes.* 1987;11 Suppl 3:179–83; Caballero B, Finer N, Wurtman RJ. Plasma amino acids and insulin levels in obesity: response to carbohydrate intake and tryptophan supplements. *Metabolism.* 1988 Jul;37(7):672–6; Caballero B, Wurtman RJ. Differential effects of insulin resistance on leucine and glucose kinetics in obesity. *Metabolism.* 1991 Jan;40(1):51–8; Kerstetter J, Caballero B, O'Brien K, Wurtman R, Allen L. Mineral homeostasis in obesity: effects of euglycemic hyperinsulinemia. *Metabolism.* 1991 Jul;40(7):707–13; Caballero B. Insulin resistance in obesity. *Arch Latinoam Nutr.* 1992 Sep;42(3 Suppl):131S–136S; Lohman TG, Caballero B, Himes JH, Hunsberger S, Reid R, Stewart D, Skipper B. Body composition assessment in American Indian children. *Am J Clin Nutr.* 1999 Apr;69(4 Suppl):764S–766S; Risica PM, Ebbesson SO, Schraer CD, Nobmann ED, Caballero BH, Body fat distribution in Alaskan Eskimos of the Bering Straits region: the Alaskan Siberia Project. *Int J Obes Relat Metab Disord.* 2000 Feb;24(2):171–9.

5. René Valley-Radot. *The life of Pasteur.* Translated by R.L. Devonshire, with an introduction by Sir William Osler (Garden City, N.Y., 1926); Gerald L. Geison. *The private science of Louis Pasteur* (Princeton, 1995).

CHAPTER 6

Federal Building
Bethesda, Maryland, May 11, 1993

This malady is endemic both in summer and in winter.
Hippocrates: "On Airs, Waters, Places"

The PAHO organizers of the Mission to Cuba, Drs Restrepo and Llanos, knew that the Mission to Cuba team would need to include a Spanish-speaking neurologist with equal expertise in epidemiology and tropical diseases of the nervous system. At first, this appeared to be an impossible feat. Both disciplines are specialized fields of study. The rosters of the World Federation of Neurology contain only a small number of neuroepidemiologists and an even smaller number of experts in tropical neurology—globally, perhaps fewer than a handful of men and women met the qualifications. In 1993, I was the only Spanish-speaking Latino holding the superior rank of Branch and Laboratory Chief at the US National Institutes of Health (NIH) as Chief of the Neuroepidemiology Branch, National Institutes of Neurological Disorders and Stroke (NINDS), in Bethesda, Maryland. I was Colombian-born, like Drs Restrepo and Llanos—and a former student of Dr Llanos—and became the last person selected to complete the Mission to Cuba team.

Date: Tuesday, 5/11/1993, Federal Building, Bethesda: I received a call early this morning from Dr Helena Restrepo, a colleague from Colombia, working at the Pan-American Health Organization (PAHO) here in Washington. She informed me that it had just been announced in Geneva that there was a large epidemic of blindness in Cuba. This confirmed the information provided on January 15, 1993, by Carlos Barros from the Cuban Trade Commission at the Cuban Consulate in Montreal, Canada. Dr Restrepo asked me if I would agree to join the Mission to Cuba as neurologist and epidemiologist. Of course, I accepted enthusiastically to participate in the Mission. "But," I told Dr Restrepo, "there is only one problem: Cuba is off-limits for me." I reminded her that as a full-time US Federal Government employee and a member of the Uniformed

Cuban Blindness. DOI: http://dx.doi.org/10.1016/B978-0-12-804083-6.00006-0
35

Services with the rank of Commander in the United States Public Health Service (USPHS), I was not authorized to travel to Cuba.

For official business, I traveled with the distinctive burgundy-colored official US passport displaying the diplomatic statement: "The Bearer is Abroad on an Official Assignment for the Government of The United States of America." Permission to travel to Cuba would require clearance at a very high level, perhaps from Admiral Novello herself, the US Surgeon General of the USPHS. "*No te preocupes* [don't worry]," Dr Restrepo answered, "I'll take care of all the arrangements."

It is now 09:00 hours. I am officially part of the Mission to Cuba team. For me, like for most in the team, this will be a trip to an unknown destination.

★★★

An urgent fax arrived at my office on Wednesday May 12, 1993, bearing the seal of the PAHO—a map of the Americas with the legend *Pro Salute Novi Mundi*, "For the Health of the New World"—and the seal of the World Health Organization (WHO)—the United Nation's North Pole view of the *mappa mundi* crossed by the medical caduceus and the two fragile olive branches of a peace wreath. The message was signed by Dr Helena Restrepo formally requesting my participation as a consultant in the Mission to Cuba; the objectives of the trip were also clearly defined: "This work will entail the following:

Review and analyze available information on the clinical and epidemiological aspects and possible causative factors of the ongoing epidemic of optic neuritis in Cuba.

Recommend further investigations needed, if any, to determine the underlying cause, and to successfully control the epidemic.

Assist in the setting-up of a proper monitoring scheme for the epidemic.

Propose mechanisms for further collaboration and coordination of work to control the epidemic."

Two days later, on Friday May 14, 1993, Linda Vogel, the Deputy Director of the Office of International Health, Department of Health and Human Services (DHHS), called and informed me that the Department of State, Office of Cuban Affairs (Cuba Desk), at the request of the WHO had authorized my travel to the island on a humanitarian basis.

However, since the US Department of the Treasury enforced the economic embargo against Cuba, it was mandatory to obtain their clearance before the trip. The PAHO was already in contact with the US Treasury, and no problems were anticipated. Finally, Linda Vogel told me that a group of ophthalmologists from a nonprofit foundation called Project Orbis, based in New York, would also be traveling to Cuba on similar dates.

A call to the Parklawn Building in Rockville, Maryland, headquarters of the USPHS, confirmed that travel clearance had, indeed, been granted. With this information at hand, I then asked Patricia Walsh, the able and efficient travel specialist of my office, to begin the lengthy process of filing the NIH travel order for Washington–Havana–Washington. The fact that this journey was sponsored by the PAHO and therefore classified as NETG (No Expense To Government) reduced the bureaucratic process only slightly. She would need to take the travel order in person from office to office. In the past, she had often secured from her desk odd-hour train connections, flawless hotel reservations, and complicated airline schedules to remote sites where research was taking place. Travel to Cuba, however, was something else.

I spent most of next day, Thursday May 13, at the NIH library reviewing scientific articles on epidemic blindness. A computer search of the National Library of Medicine's database MEDLARS revealed no recent articles of relevance. There were mainly old references in the BACK-76 files covering articles published before 1976. TOXLINE, the toxicology database, showed only a few papers on methyl alcohol and several individual case reports of toxicity to the eye from a number of pharmaceutical drugs. I selected the most relevant titles and walked along the shelves of the silent library preparing the articles for photocopying. As an afterthought, just before leaving I searched the Internet again for international news clips using the key words "Cuba" and "epidemic," and obtained a few interesting entries from Reuters.

Back in my office, I found a cryptic message from "Commander Kevin Tonat, aide-de-camp to Admiral Young of Emergency Preparedness, Ref. Cuba." I called Admiral Young, who briefed me on the facts learned about the epidemic and we discussed at length its possible causes. We both agreed that it was mandatory to examine the patients before reaching a conclusion. He asked to be informed at once if a transmissible agent could be suspected to be the cause of the epidemic disease. I was to communicate the news through the United States Interests Section in Havana[1]—the old US Embassy Office officially closed since 1961 at the height of the Cold War and reopened on July 20, 2015, after 54 years of closure. Upon my warning, biologic barriers would be raised immediately to block the entry of any pathogenic organism into the continental United States. I could imagine the transportation nightmare that would result from an imposed quarantine on every human being arriving in the United States from the Caribbean.

Toward the end of the day I received an unexpected visitor. The visitor's card printed in fine Palmer cursive read:

> *Pablo R. Rodríguez*
>
> *2nd. Secretary, Cuban Interest Section,*
>
> *2630 16th St, NW Washington, DC 20009*

Mr Rodríguez was a career diplomat and his demeanor revealed his long experience in "the art of meeting situations without arousing antagonism."[2] He had been in Washington for many years, almost since 1977 when the Carter administration allowed the establishment of diplomatic "interests sections" in Havana for the United States and in Washington for Cuba.[3] The reason for his visit was to bring in person my visa for entering Cuba—a folded page of official green paper properly sealed to keep it separate from the passport in order to avoid a permanent record of the visit to Cuba.

Mr Rodríguez talked about the problems faced on the island by physicians and scientists trying to obtain scientific information. Computer search of bibliographic sources was impossible because there was no international telephone service; subscriptions to scientific journals had to be suspended because of lack of funds; books were too expensive; and even photocopies were out of the question because of shortage of paper. He suggested that it would be helpful to bring copies of relevant articles for the Cuban colleagues.

Then, abruptly changing from his excellent English into Caribbean Spanish—a fast and loud Seville intonation that drops the final *s*'s and sounds out the *r*'s like *l*'s—he proceeded to give me the second briefing of the day on the Cuban epidemic.

The overall facts presented by my two sources—Admiral Young and Mr Rodríguez—agreed in their general outline. However, when Rodríguez concluded his statement, he stood up to leave, and almost at the door he told me in a confidential tone, "We are concerned about the possibility that this disease could be due to covert action from the CIA."

Then, suddenly exasperated, he said loudly, "You know, the CIA tried to assassinate Castro several times, and they were responsible for introducing epidemics of porcine fever and dengue in Cuba."[4]

And then, he added emphatically, "¡*Eso está demostrado!* This has been proven!" and he struck his fist on the open palm of his hand, like a boxer.

★★★

From previous experiences in epidemiologic fieldwork, I had learned the value of collecting blood samples from patients. Even under conditions of poor refrigeration and problems with transportation, these samples had proven to be invaluable. Often, researchers are given a single vanishing opportunity to secure the samples that may hold the answer to a mysterious illness. Therefore, I carefully prepared large Styrofoam supply boxes that included hypodermic needles, vacutainers, elements for venipuncture, alcohol pads, plastic tubes for freezing serum samples, plastic bags, notebooks, pens, labels, and a number of small items that, as my experience had shown, cannot be taken for granted overseas. My secretary added soap and cookies and suggested that toilet paper should be included. I checked with Dr Helena Restrepo, who reassured me that the hotel in Havana would have soap, toilet paper, and appropriate food. Green-and-yellow customs forms for "Plant Quarantine Material" from the US Department of Agriculture granting permit to import plants or plant products, and bright red stickers with a triangle of warning circles from the Centers for Disease Control and Prevention (CDC) "to import or transport agents or vectors of human disease" were obtained. The red stickers on the icebox would be a clear sign of biologic danger. The order "Do NOT Open in Transit" and the official passport would ensure prompt customs clearance in Cuba and Miami.

ENDNOTES

1. Office of the United States Interests Section in Havana, Cuba: See www.havana.usint. gov.
2. Definition of diplomacy, according to Webster's.
3. Thomas G. Paterson. *Contesting Castro* (New York, 1994).
4. The official *Nuevo Atlas Nacional de Cuba* (España 1989), lists under title XXIV: *Historia y Revolución* (p XXIV. 3.2) the following entries: "May 7, 1971: Onset of African Swine Fever epidemic, introduced in Cuba by a CIA agent, resulting in elimination of half-million animals." "July 1977: Onset of dengue fever (virus type 1) epidemic introduced by the CIA." "May 1981: Onset of epidemic of dengue hemorrhagic fever, introduced in Cuba by CIA agents."

CHAPTER 7

Neuroepidemiology, NIH
Bethesda, Maryland, May 14, 1993

Why give light to men in grief?
Job, 3, 20

On Friday May 14, 1993, at 17:00 hours, just at the closing of the Federal day, all boxes had been packed, and I had received my official passport and the travel order signed by the director of the National Institute of Neurological Disorders and Stroke (NINDS), Dr Murray Goldstein.

My journey to Cuba was about to begin.

Only 3 years earlier—in September 1990—I had accepted the position of Director of the Neuroepidemiology Branch of the NINDS, part of the National Institutes of Health (NIH), the world-renowned medical research institute of the United States. I considered this appointment a crowning achievement in my career as an academic neurologist. It was a moment of great triumph, coming at a time of terrible grief.

On July 22, 1990, my daughter Natalia Isabel had fallen ill suddenly. Natalia Isabel and her older brother Gustavo were spending 6 weeks of their summer vacation visiting my mother and father at their home in Bogotá, Colombia. It was, we felt, a good opportunity for the children to visit our large family and get reacquainted with their roots, as well as a chance to improve their conversational Spanish by interacting with the large clan of cousins. Gustavo and Natalia also planned to spend some days at my father's small farm to learn that the hanging fruits in the tropical trees have aromas and sweet flavors that can never be found in the produce displayed in the supermarkets in the United States; to wake up early to help my father milk the cows; to pet the newborn calves; and to gather eggs in the henhouse and hold yellow fluffy chicks in the cradle of their hands. My wife Lydia and I had stayed behind at our home in Lubbock, Texas. We planned to travel to Colombia in the first week of August to pick up our children in time for the return to school.

Cuban Blindness. DOI: http://dx.doi.org/10.1016/B978-0-12-804083-6.00007-2

It was not to be. Early one morning the telephone rang. Gustavo, clearly agitated, said, "Daddy! I cannot wake Natalia up. I looked at her pupils, and the right one is larger than the left."

A neurologist's son, he sometimes accompanied me to the hospital on rare Sunday rounds and had seen me checking patients' pupils. I froze at his words because the diagnosis was perfectly clear in my mind: a cerebral hemorrhage. This was an old foe that I had faced countless times, the most lethal form of stroke. How many times I had tried to convey gently to the anxious relatives the hopelessness of the situation, explaining to them that the white smear on the computed tomography scan of the brain was slowly disconnecting vital functions and that there was nothing we could do—no medicine to use or no heroic surgery to attempt.

I talked to my father who had already recognized the seriousness of the situation and had called an ambulance. I told him to call two of my former teachers and dear friends, a neurosurgeon and a neurologist, and my father assured me that they were both on their way and they would be taking Natalia Isabel to the Marly Clinic.

Natalia Isabel was always a healthy child, but about 1 year earlier, during the routine physical examination at school, her pediatrician had found that her blood pressure was elevated. Hypertension in children is unusual, and specialists conducted numerous laboratory and imaging tests to uncover the cause of her high blood pressure but to no avail. One small tablet of antihypertensive medication controlled her blood pressure, and she accepted the responsibility of taking her tablet daily in the morning. Gustavo and my mother confirmed that Natalia had continued taking her little tablet every single day.

The next day, after a dreadful journey from Dallas to Miami and then to Bogotá, Lydia and I were at Natalia's hospital bedside. She had lapsed into a coma and was on a respirator. We reviewed the brain images with her doctors—our teachers—and it was terribly clear that my suspicions were correct: a ruptured blood vessel had caused a hemorrhage in her brain. We sat for long hours, holding her still-warm hand, telling her gently of our love for her.

She died, the following day, July 24, 1990, without regaining consciousness. She was 13 years old.

It was a devastating loss, from which I did not expect to recover. I felt I could not go on. In mourning, I decided to leave Texas, to go into seclusion, to stop seeing patients, to stop experiencing their suffering. For the past 15 years I had been working as a clinical neurologist. Now I felt I was finished. I must seek a new beginning, in a different city, with other challenges and unknown problems. Six months earlier, in Bethesda, a colleague suggested I allow my name to be put in for the vacant NIH position. At that time I was reluctant to leave the clinic, but in September, 6 weeks after the death of my only daughter, I accepted the recommendation. I was appointed in the US Public Health Service with the rank of Commander. Lydia and I moved with the boys, Gustavo and our youngest son Andrés, to Washington.

The home quarters of the Román family was a Victorian house at No. 4 West Drive on the NIH campus, shielded by green lawns and centenarian trees from the noise of the city and the traffic of Washington's beltway. The house had previously been home to C. Everett Koop, the former Surgeon General.[1] I learned after we moved in that Dr Koop and I belong to the secret-sorrow, black-armband brotherhood of fathers who had lost a child in the prime of their lives. A portrait of Koop's son David, who had died at age 20 in a 1968 rock-climbing accident, hung over the fireplace, where I decided to place Natalia's portrait as well.

The words *neuro* and *epidemiology* on my office door, on the seventh floor of the Federal Building in downtown Bethesda, implied that we dealt with epidemics of nervous system diseases. But rather than fighting epidemics, we were more occupied with keeping track of published papers on the frequency of common diseases such as stroke, multiple sclerosis, dementia, and Parkinson disease. Nonetheless, my major interest in science was tropical neurology—exotic diseases that occur in the jungles, mountains, and deserts of the tropical world. At the Neuroepidemiology Branch I swiftly established a close-knit international research network.

Life's road had brought me to the NIH, but I remained the same Gustavo Román, a Colombian physician born in Santafé de Bogotá, a city of cold nights with rain water that reflected the street light on the black pavement, a city filled by day with luminous shades of green, and a city perched high in the flat lands of the Andes and surrounded by giant blue-slate mountains.[2]

I was the oldest of a family of six boys and three girls. We grew up in a corner house of a middle-class housing project called "Los Alcázares," a name that my dictionary translates, strangely, as "the royal palace." My mother who had a gift for poetry was educated to be a schoolteacher, but her real job was caring for her large family. My father was a veterinarian, who taught me the love for the biologic sciences and the importance of careful observation as a tool for clinical diagnosis.

In 1971, I graduated in medicine from the National University of Colombia in Bogotá and completed a 1-year rotating internship at the "San Juan de Dios" University Hospital. This was originally a turn-of-the-century charity hospital, with large pavilions in the French style, famed for its rich clinical material—an euphemism for the interminable stream of human misery knocking at its doors. We also rotated at the maternity and pediatric hospitals called "Materno-Infantil" and "La Misericordia."

After my internship I performed 1 year of social service at a sanatorium for tuberculosis run by Catholic nuns. The hospital brought to mind the clinic in Thomas Mann's *Magic Mountain*. Many patients had tuberculosis resistant to multiple drugs, forcing their doctors to turn back the medical clock to deflation of lungs and thoracotomies, which were the usual treatments in the days before streptomycin.

Even as a medical student, my interests had focused on neurology and psychiatry. I began teaching neuroanatomy at the National University of Colombia and worked as an intern at a psychiatric clinic. Professor Ignacio Vergara introduced me to the peculiar patterns of neurology in the tropics. Under his direction and that of Professors Jaime Saravia and Gabriel Toro, a classmate and I reviewed hundreds of medical records and autopsy protocols of adult patients with infections of the nervous system. This was my first experience in clinical research, which eventually resulted in the publication of a series of articles on meningitis in adults, brain abscess, tuberculomas, and cysticercosis.[3]

The classmate was Lydia Navarro. While spending unending hours in the endless review of medical records at the hospital, we fell in love. True love, we discovered, was working and living together. We got married at the University chapel on Bastille's Day, July 14, 1972.

Shortly after our marriage, with a scholarship from the French Ministry of Foreign Affairs, we traveled to France and spent that summer in Vichy studying medical French in preparation for our life in Paris.

That autumn, I began studying neurology at the Salpêtrière Hospital—the birthplace of neurology. Lydia began her specialty in infectious diseases and epidemiology at the Pasteur Institute. We lived in a small apartment, shared with a couple of exiled Chilean students, at No. 4 Rue Saint Martin, near the Saint-Jacques tower, the starting point for all pilgrims traveling by foot to visit the tomb of the apostle Santiago in Compostela, Spain.[4] On a winter afternoon in December 1973 our son Gustavo was born at the maternity ward of the Pitié Hospital, in the heart of Paris.

After two-and-a-half years in Paris, Lydia and I passed the US Foreign Medical Graduates Examination (ECFMG) and decided to come to the United States to complete our training. I was offered a position as neurology resident at the University of Vermont in Burlington, Vermont, and my wife was admitted as an intern in Pathology. This was a productive time, and in my third year I was appointed Chief Resident in Neurology and 1 year later began a Fellowship in Clinical Neurophysiology and Neuropathology. On a bright summer day in 1977, at noon, our daughter Natalia Isabel was born at the Mary Fletcher Hospital in Burlington, Vermont.

A year later we returned to Colombia after an absence of 8 years. Having completed our training in France and the United States, back in Bogotá I was appointed assistant professor of neurology at my Alma Mater and worked at a nonprofit foundation for the advancement of biomedical sciences. In 1979, almost 20 years before the birth of the Internet, we obtained for Colombia satellite linkage with the MEDLARS-MEDLINE database of the National Library of Medicine, situated in the Bethesda campus of the NIH. Perhaps, that first contact with the NIH planted in me the desire to work one day at that prestigious institution.

During those years in Colombia I wrote a neurology textbook, *Neurología Práctica*,[5] that was popular among students and was awarded the National Medical Prize by the Colombian Academy of Medicine in 1980. Also, Lydia and I wrote with Professor Gabriel Toro one of the few existing textbooks on tropical neurology, *Neurología Tropical*.[6]

In 1982, Lydia and I traveled to the port town of Tumaco, in the Colombian lowlands of the Pacific coast, to study an outbreak of a strange neurologic disease we named "tropical spastic paraparesis" (TSP).[7-11] Its manifestations, including paralysis of the legs, were somewhat similar to those of multiple sclerosis. TSP affected older black persons from this region and attracted the attention of Drs Bruce Schoenberg and Peter Spencer from the United States. Dr Schoenberg was then Chief of Neuroepidemiology at the NIH. I worked closely with him in Tumaco, but I never imagined that one day I would succeed him at the NIH.

Restless to continue my academic growth, in 1983 I joined the faculty at Texas Tech University School of Medicine in Lubbock, Texas. Once again, Lydia and I left our home, family, and friends to a voluntary exile. In 1984, our youngest son Andrés Santiago—a true native Texan—was born.

In 1986, along with Drs Bruce Schoenberg and Peter Spencer, we visited Mahé in the Seychelles Islands of the Indian Ocean to investigate an epidemic outbreak of TSP. This study helped demonstrate that TSP is caused by HTLV-I,[12] the first of the human retroviruses, a family of viruses that includes HIV, the cause of AIDS.

During our years in Lubbock I served as Acting Chairman of Texas Tech's combined department of Neurology and Neurosurgery. Finally, in September of 1988, at age 42, I became Tenured Professor of Neurology.

I went back to Colombia for a brief visit in 1992 to attend my mother's funeral. She had never recovered from the grief of losing her first granddaughter "on her shift," as she used to say. My father was then in his late 80s and was having health problems, but he continued to be a sharp observer of people and animals. He was very proud of the professional achievements of his children. I told him that the lessons I had learned from him had been invaluable in my career and would continue to be useful while facing future challenges.

Despite all my preparation, though, the truth is that after the study of the outbreaks of TSP from HTLV-I infection, during my 3 years at the NIH the opportunity to study another epidemic of neurologic disease had never arrived—not until March 23, 1993, when I received a telephone call from Carlos Barros, Cuba's business representative in Montreal, Canada. Like all members of the consular service, Barros had received from Havana the request to find information on "optic neuritis." To my astonishment, he informed me of "an epidemic in Cuba with hundreds of cases of retrobulbar optic neuritis…."

ENDNOTES

1. Gregg Easterbrook. *Surgeon Koop: Medicine and the politics of change* (Knoxville, 1991).
2. Santafé de Bogotá was founded on August 6, 1538 by Gonzalo Jiménez de Quesada. The view of the verdant planes and the cool winds from the mountains, after the steaming humidity, insects, and plagues of the jungle during the ascent of the Magdalena River valley, inspired Don Juan de Castellanos the following verse: Good land! Good land! Our suffering is over!
3. I. Vergara, J. Saravia, G. Toro, G. Román, L.I. Navarro, Meningitis del adulto: Revisión clínica y patológica de 400 casos. *Revista de la Facultad de Medicina, Universidad Nacional de Colombia* 1971; 37:321–379.
4. El Camino de Santiago: *Guía del Peregrino* (Madrid, 1985); Camino de Santiago Web Links at: http://www.netcomuk.co.uk/~downs/santiago.html.
5. Gustavo Román. *Neurología Práctica. Principios de fisiopatología, clínica y terapéutica neurológicas* (Bogotá, Colombia, 1982).
6. Gabriel Toro, Gustavo Román, Lydia Isabel Navarro de Román. *Neurología Tropical: Aspectos Neuropatológicos de la Medicina Tropical* (Bogotá, Colombia, 1983).
7. Tropical spastic paraparesis or TSP is a tropical disease characterized by weakness of the legs. A number of these patients are infected by the human T-lymphotropic virus type I (HTLV-I), the first human retrovirus to be discovered. This same agent causes adult T-cell leukemia/lymphoma, a hematologic malignancy prevalent in Japan and the Caribbean.

8. Román GC, Román LN, Spencer PS, Schoenberg BS. An outbreak of spastic parapa-resis along the southern Pacific coast of Colombia: Clinical and epidemiological fea-tures. *Ann Neurol* 1983;14:152A.
9. Román GC, Roman LN. Tropical Spastic Paraparesis in the Pacific Lowlands of Colombia. *REPORT TO THE INTERNATIONAL RESEARCH GROUP FOR THE STUDY OF TROPICAL SPASTIC PARAPARESIS.* National Institutes of Health, Bethesda, Maryland, April 16, 1984.
10. Román GC, Román LN, Spencer PS, Schoenberg BS. Tropical spastic paraparesis: Neuroepidemiological study in Colombia. *Ann Neurol* 1985;17: 361-365.
11. Román GC, Román LN. Tropical spastic paraparesis in the Pacific lowlands of Colombia: A clinical and laboratory study of 50 cases and review of the literature. *J Neurol Sci* 1988;87:121-138.
12. Gustavo C Román, Jean-Claude Vernant, Mitsuhiro Osame (editors). *HTLV-I and the nervous system* (New York, 1989).

CHAPTER 8

Cuba: A Dream Island
May 14, 1993

*…find your way to Cuba on the map: a long, green alligator,
with eyes of stone and water.*
Nicolás Guillén

In images from weather surveillance satellites Cuba looks like a giant caiman sleeping half-submerged in water, blocking the entrance to the Gulf
of Mexico, and lying with its back turned to the Straits of Florida, the
Florida Keys, and the Bahamas in the Atlantic Ocean. The slender tail
points west, separating the Gulf of Mexico from the Caribbean, cradled on
the curve of the island's southern flank. At the eastern end, the mountains
of Sierra Maestra and Baracoa appear as the caiman's bony head, watching
the islands of Jamaica and, just beyond the Windward Passage, Hispaniola.

Cuba's strategic location and vast size—777 miles (1250 km) long—
made this island one of the most important early settlements of the
Spanish crown in the New World and the last colony to become independent. Cuba remained under Spain's gold-and-crimson flag until 1898,
when the United States approved Cuba's independence at the conclusion
of the Spanish–American War.

Cuba is *una isla de ensueño*, a dream island. Even today, Christopher
Columbus' ship log entry from Sunday, October 28, 1492, clearly conveys
the explorer's delight at his discovery:

*This island is the most beautiful that eyes have ever seen. It has such marvelous
beauty that it surpasses all others in charms and graces … I have been overwhelmed at this sight of so much beauty that I have not known how to relate it.*
La Santa María's log[1]

Then, as now, forests of tall Royal palms—Cuba's national tree—adorn
the landscape. Legend has it that these were native Taino Indian princesses—
tall, slender and glamorous, with fine, flexible waists, and long hair stream-
ing in the breeze—always in small groups talking to one another. Elsewhere,
the curved profile of the Belly palms, or *barrigonas*, suggests that these were
the Taino wives, proudly displaying their pregnant bellies. Very few elements

Cuban Blindness. DOI: http://dx.doi.org/10.1016/B978-0-12-804083-6.00008-4

survived the extinction of the Taino Indians at the hands of the conquering white men. Among the remnants are the legends surrounding the splendid palm trees, the thick, long, dark hair of Cuban women proclaiming the mixture of Amerindian and Spanish blood, and, perhaps more pervasively, a few remaining Native American words—*barbecue, canoe, cacique, cassava, hammock,* and *tobacco*.

The tobacco tradition in Cuba extends back to pre-Columbian times. Columbus' sailors learned from the Taino how to "smoke perfumed herbs ignited with live coals which they carried around with them."[2]

Along with tobacco, Cuba's name is flavored by the sweetness of sugar. Sugar cane (*Saccharum officinarum* and other species) was domesticated in New Guinea 12,000 years ago and was dispersed to Asia and Africa. From North Africa, the Moors brought it to the Iberian peninsula, and in the fifteenth century the Portuguese planted it in the islands of Madeira and São Tomé, as did the Spaniards in the Canary Islands. On his second voyage, Columbus brought sugar cane to the New World. It was first planted in December 1493 in Santo Domingo, Hispaniola Island.[3] Soon, lush sugar cane plantations grew on the islands of Cuba, Puerto Rico, and Jamaica, to eventually become the economic foundation of the entire Caribbean region. Sugar-based economies persisted until the end of the twentieth century when corn syrup largely replaced cane sugar in North America.

The first Spanish settlers arrived on Cuba in 1511, more than 100 years before the Mayflower pilgrims arrived in North America. The town of Baracoa was founded in 1512, followed shortly thereafter by Bayamo, Sancti Spiritus, Camagüey, Santiago de Cuba, and, in 1519, San Cristobal de la Habana, called Havana in English.[2]

Cuba became the departure point for the conquest of the New World. The explorer Hernán Cortés went to Mexico from Santiago de Cuba. The gold of the Americas passed through Cuba as the cargo of galleons on its way to the Old Continent. Juan Ponce de León, the discoverer of Florida, died here.

Cuba was called "the Spanish pearl" at the heart of the Caribbean and became a coveted prize for the English, French, Dutch, and Portuguese pirates, privateers, corsairs, and buccaneers who roamed the Caribbean. Many attempts at conquest were made. Cuba resisted with the might of its fortifications. Havana's official coat-of-arms proudly displays the key to the New World, and three castles—El Morro, La Fuerza, and La Punta—continue to guard the bay still today.

Havana—never before defeated—was briefly taken and occupied by the British in 1762 after the victory of Lord Albemarle. The defeat opened the island to English merchants and slave traders. The productivity of sugar plantations worked by African slave labor increased substantially. One year later, after the Peace of Paris in 1763, England returned Cuba to the Spanish crown, but the sugar mills had undergone a major transformation.

By the 1790s Cuba was exporting over 30,000 tons of sugar annually. Sugar cane cultivation naturally led to the production of rum by fermentation and distillation of the sweet molasses. Eventually, the entire economy of the island came to depend almost completely on sugar exports to the United States until the collapse of the sugar trade—first with the end of slavery in 1840 and, more recently, in 1960 when the Soviet Union took over the US sugar market.

Cuba was once a favorite vacation spot for North Americans. In the 1950s some 300,000 tourists visited the island each year. Travel to Cuba was by ferry from Key West or by Pan American flight from Miami. However, for more than 30 years—since January 1961 when the United States broke diplomatic relations with Cuba, and later on with the implementation of the US economic embargo—it has been problematic for US residents to travel to Cuba, and conversely, an almost impossible feat for Cubans to leave the island.

During the 1990s it became far more difficult to schedule a trip from Miami to Havana—a short 35-min flight—than to reach the most distant points of Africa, Asia, or the Indian Ocean. Cuba, the largest island of the Caribbean, simply ceased to exist on the computers of US airlines, hotels, and travel services.

At the time of our trip to Cuba, there were weekly flights to Havana from Mexico City, Santo Domingo, Europe, and Latin America, but only two companies, Marazul and CBT, chartered flights from Miami to Havana. The former leased planes and crews from Haiti Trans Air and the latter from Lloyd Aéreo Boliviano—among the least known airlines in the world.

Years of isolation eventually shrouded the island in a veil of mystery. The Cuba we were going to visit, the only Communist country in the Americas, had a justified reputation for seclusion.

The Mission to Cuba team had been completed and comprised the following scientists: Dr Guillermo Llanos, epidemiologist from the PAHO in Washington DC; Dr Juan Carlos Silva, Director of the Blindness Prevention Program in the Americas from Bogotá, Colombia;

Dr Benjamin Caballero, nutrition expert, Director of the Division of Human Nutrition at the Johns Hopkins School of Public Health in Baltimore, Maryland; Dr Rafael Muci-Mendoza, Professor of Internal Medicine and Neuroophthalmology at the Central University, Caracas, Venezuela; and, Dr Gustavo Román, neurologist and neuroepidemiologist, Chief of the Neuroepidemiology Branch of the National Institutes of Health (NIH) in Bethesda, Maryland.

The next stop would be the international airport in Miami, Florida.

ENDNOTES

1. Zvi Dor-Ner. *Columbus and the age of discovery* (New York, 1992). There are several versions of Columbus' log: Clements R. Markham (Trans.) *The journal of Christopher Columbus (during his first voyage, 1492–93) and documents relating to the voyages of John Cabot and Gaspar Corte Real* (London: Hakluyt Society 1893); Excerpts from Christopher Columbus' Log, 1492 A.D. https://franciscan-archive.org/columbus/index.html "This island is the most beautiful that I have yet seen, the trees in great number, flourishing and lofty; the land is higher than the other islands, and exhibits an eminence, which though it cannot be called a mountain, yet adds a beauty to its appearance, and gives an indication of streams of water in the interior. From this part toward the northeast is an extensive bay with many large and thick groves. This is so beautiful a place, as well as the neighboring regions, that I know not in which course to proceed first; my eyes are never tired with viewing such delightful verdure, and of a species so new and dissimilar to that of our country." The Log of Christopher Columbus' First Voyage to America in the Year 1492 Christopher Columbus 1451–1506 Bartolome de Las Casas c. 1490–1558 Original Source: Christopher Columbus, "Journal of the First Voyage of Columbus," in Julius E. Olson and Edward Gaylord Bourne, eds., *The Northmen, Columbus and Cabot, 985–1503, Original Narratives of Early American History*. New York: Charles Scribner's Sons, 1906. Early Americas Digital Archive. http://mith.umd.edu/eada/html/display.php?docs=columbus_journal.xml.
Sunday, 28th of October
"I went thence in search of the island of Cuba on a S.S.W. course, making for the nearest point of it, and entered a very beautiful river without danger of sunken rocks or other impediments. All the coast was clear of dangers up to the shore. The mouth of the river was 12 *brazas* across, and it is wide enough for a vessel to beat in. I anchored about a lombard-shot inside." The Admiral says that he never beheld such a beautiful place, with trees bordering the river, handsome, green, and different from ours, having fruits and flowers each one according to its nature. There are many birds, which sing very sweetly. There are a great number of palm trees of a different kind from those in Guinea and from outs, of a middling height, the trunks without that covering, and the leaves very large, with which they thatch their houses. The country is very level. The Admiral jumped into his boat and went on shore. He came to two houses, which he believed to belong to fishermen who had fled from fear. In one of them he found a kind of dog that never barks, and in both there were nets of palm-fiber and cordage,

as well as horn fish-hooks, bone harpoons, and other apparatus for fishing, and several hearths. He believed that many people lived together in one house. He gave orders that nothing in the houses should be touched, and so it was done. The herbage was as thick as in Andalusia during April and May. He found much purslane and wild amaranth. He returned to the boat and went up the river for some distance, and he says it was great pleasure to see the bright verdure, and the birds, which he could not leave to go back. He says that this island is the most beautiful that eyes have seen, full of good harbors and deep rivers, and the sea appeared as if it never rose; for the herbage on the beach nearly reached the waves, which does not happen where the sea is rough (Up to that time they had not experienced a rough sea among all those islands.). He says that the island is full of very beautiful mountains, although they are not very extensive as regards length, but high; and all the country is high like Sicily. It is abundantly supplied with water, as they gathered from the Indians they had taken with them from the island of Guanahani.

2. A. Gerald Gravette. *Cuba: Official Guide* (London, 1988); Daniel J. Boorstin. *The discoverers* (New York, 1985); Germán Arciniegas. *Biografía del Caribe* (Buenos Aires, 1973); Juliet Barclay. *Havana: Portrait of a city* (London, 1993); Hugh Thomas. *Cuba, c. 1750–c. 1860.* In, Leslie Behtell (ed.). *Cuba—A short history* (Cambridge, 1993).

3. Sidney W. Mintz. *Pleasure, profit, and satiation.* In, Herman J. Viola and Carolyn Margolis (Editors). *Seeds of change* (Washington, 1991).

CHAPTER 9

A Farcical Journey

On this journey of journeys, solitude found solidarity, I turned into we.
Eduardo Galeano

Date: Saturday, 5/15/1993, Miami International Airport, Midnight: The PAHO's travel office was able to secure airplane seats with Marazul CBT Charters for all the team members of the Mission to Cuba despite the very limited seating and bookings usually made months in advance by Cuban nationals that obtain permission to visit relatives in Miami or by older Cuban-Americans who applied for Cuban visas to visit the island of their dreams one last time.

Our instructions were clear:

> *You should arrive at Miami International Airport the night before departure. A room has been reserved for you at the airport hotel. Flight departs from Miami at 9:00 a.m. Airline has requested that you be at Marazul ticket counter concourse B second floor at 5 a.m. Confirmation number is 0512120. You should have your passport and Cuban visa ready. Payment in cash of $225.00 (round trip) plus taxes is requested.*

The last members of our team arrived at the Miami International Airport from Washington National and Baltimore-Washington. Room reservations were honored for late arrivals at the airport hotel. At 11:30 p.m.—amid the traffic of travelers and luggage—we met briefly with Dr Llanos in the hotel lobby, which was decorated in greens, oranges, and purples and displayed a loud TV set and gaudy mirrors. We agreed to reconvene for checkout at 04:30 a.m. the next day. We had just a few hours to sleep. With the exception of Dr Llanos, we were all traveling to Cuba for the first time. With a sense of foreboding, I begin this journey.

Date, Sunday, May 16, 1993 Miami International Airport, 05:00 Hour: There was a market-place air of excitement at dawn at the Marazul ticket counter of the Miami International Airport. Travel for Latinos is an important family affair. *Al viajero, ayudarle a salir*, help the traveler depart, goes the saying from a time when travel meant leaving

Cuban Blindness. DOI: http://dx.doi.org/10.1016/B978-0-12-804083-6.00009-6

family and friends, perhaps forever, to settle in remote corners of the Spanish colonial empire. Travel also provides an outlet to express the powerful sense of family so strongly present in Latinos in general and Cubans in particular. Married sons, daughters-in-law, brothers, sisters, children, and grandchildren were here to see mamá's departure for Cuba. Her huge suitcases were weighed and their contents reviewed once and again to fit the allowed weight limit.

—Mamá, leave the bottle of shampoo. It's too heavy.
—But I have nothing for Carmen.
—Yes, give her the blouse.

★★★

Large transparent plastic bags, overflowing with medications, were being carried on board as hand luggage. They contained vitamins of all colors and brands, blood pressure pills, antibiotics, heart medications, aspirin, Tylenol, Nuprin, pills for menstrual cramps, tampons, Kleenex tissues, toilet paper, cortisone cream, douches, yeast infection medications, fish oil capsules, and protein pills and amino acids from health food stores—essential items not easily available or not available at all in Cuba. The bags also held items deemed indispensable by the fastidiously clean Cubans, such as deodorant, soap, disposable razor, shaving cream, shampoo, hair spray, after-shave lotion, small bottles of perfume, tooth brush and dentifrice, skin care cream, makeup, eyeliner, eye shadow, lipstick, and nail polish. These were rare goods in Cuba, treasures brought by travelers from the neighboring country to the north.

Cuban nationals returning to their homeland were easy to distinguish from other travelers. Respectable old ladies arrived at the airport wearing three dresses of different colors, one over the other, and large, wide-brimmed hats glittering with stitched-on costume jewelry, and their earrings, bracelets, and rings clinking at every step. Older Cuban men wore several shirts and several pairs of jeans one over the other. Serious-faced young men looked absurd under four or five felt hats, stacked one on top of another, or a pile of New York Yankees baseball caps. A Cuban matron walked carefully, solemnly wearing clown-size Adidas high-tops, a gift for a beloved grandson. Since items worn by travelers were not included in the weight limitations imposed by the airline, Cubans maximized in this way the amount of gifts brought into the island.

★★★

At 6:00 a.m. the lights in the Marazul airline counter were turned on, and a bilingual leaflet signed by the US Department of the Treasury, Office of Foreign Assets Control, was distributed to all travelers to ensure compliance with the economic embargo law. It read:

Announcement:

Ladies and gentlemen, there is an economic embargo against Cuba!

"Travelers are allowed $100 per day for authorized travel-related expenses and up to $100 to buy goods in Cuba. For relatives in the island, $300 for support per household in any 3-month period; for emigration from Cuba, a one-time payment of $500 per recipient. No one is authorized to carry family remittances on behalf of people outside his own household. Moreover, passengers who are traveling as fully hosted or sponsored visitors to Cuba are not permitted to carry any money for living expenses in Cuba or for the purchase of merchandise there. Be sure that you are in compliance with US law before you travel."

At the counter we were identified as "the United Nations team," and we were asked to provide passports and to return in 1 h. We sat down at the nearby coffee shop for a quick breakfast, and while we waited, the team of representatives of ORBIS International arrived.

ORBIS is a nonprofit organization, born in Houston, Texas, but based in New York, dedicated to the global fight against blindness. Since 1982, its team has circled the world three times aboard a hospital airplane, bringing ophthalmologic expertise and teaching to more than 10,000 doctors in 60 countries. In July 1991, its white Douglas DC-8 jet landed in Havana in response to an invitation extended by the Cuban Ministry of Public Health and the Cuban ophthalmology society. The ORBIS team stayed 3 weeks, concentrating on glaucoma, operating on patients using state-of-the-art laser and microsurgical techniques. There was a fruitful interchange of information and knowledge, providing Cuban ophthalmologists an opportunity to present their results of a novel treatment for retinitis pigmentosa. The surgical ORBIS team was not told at the time about the incipient outbreak of the epidemic of blindness that was brewing in the island.

It was by sheer coincidence that we ran into them at the airport. Entirely independent of the Mission to Cuba effort, Fidel Castro had asked his Minister of Health to request assistance from his friends at ORBIS. Now, at dawn in Miami, introductions were passed around the table. I met the team leaders—Dr James F. Martone, MD, MPH, Medical Director of ORBIS; and Dr Alfredo A. Sadun, MD, PhD, Professor of Neuroophthalmology at the Departments of Ophthalmology and Neurological Surgery, University of Southern California and the Estelle Doheny Eye Medical Clinic in Los Angeles, California.

Dr Sadun walked with a cane. "Recovering from knee surgery," he explained. The walking stick, along with his balding head, thick mustache, and professorial demeanor, gave him an elegant old-world air of distinction. Even in casual conversation he spoke as if he were lecturing a student audience taking notes.

Born in New Orleans into an immigrant Jewish-Italian family, Dr Sadun had enjoyed a brilliant career. Educated at the Massachusetts Institute of Technology (MIT) and Albert Einstein University, he obtained his PhD degree in Neuroscience in 1976 with his thesis on the role of light in the setting of circadian rhythms.[1] Two years later he received his MD degree; he specialized in ophthalmology at the Massachusetts Eye and Ear Infirmary and in Neuroophthalmology at Boston University and Harvard Medical School, rapidly becoming a recognized expert in optic nerve problems. In 1984, at age 40, he became full professor in the Departments of Ophthalmology and Neurological Surgery at the University of Southern California.

Dr Jim Martone, also of Italian ancestry, had dark brown hair, a mustache, and a youthful demeanor that belied his long experience in international affairs. He began his education at Providence College in Rhode Island but went to Cebu City in the Philippines to begin his medical education. This early exposure to the tropics, its people, and its problems clearly defined his field of medical activities in the years to come. A classmate of Dr Sadun, he also graduated in 1978 at Albert Einstein University School of Medicine. Dr Martone trained in ophthalmology at Montefiore Hospital in New York. He practiced in Castries on the island of St. Lucia in the Caribbean, under the auspices of the International Eye Foundation. This experience in the field led him to broaden his knowledge about blindness as a global problem from epidemiology and public health viewpoints.[2] Back in Baltimore he obtained a master's degree in Public Health from the Johns Hopkins' School of Public Health and became Medical Director of ORBIS International.

Dr Martone would bring to the team his dual expertise in epidemiology and ophthalmology, and he would, in particular, impress upon the team the need for strict diagnostic criteria that would prevent "hyperdiagnosis"—the trap of including as part of the epidemic people not really affected by the disease, thereby inflating the figures and distorting the facts. Avoiding hyper-diagnosis would require finding a distinctive feature of the disease that would identify it rapidly and accurately among the long lines of people complaining of poor eye sight.

In other words, Dr Martone would insist on finding a sieve to separate the wheat from the chaff.

ENDNOTES

1. Circadian: of 1-day duration (Latin *circa* about, *dies* a day); Alfredo A. Sadun, et al. A retinohypothalamic pathway in man. Light mediation of circadian rhythms. *Brain Research* 1984; 302: 371–377.
2. Grimes MR, Scardino MA, Martone JE. Worldwide blindness. *Nursing Clinics of North America* 1992; 27(3): 807–816.

CHAPTER 10

In the Eye of the Storm

Havana…is a most extensive city, of regular plan, one of the best fortified in America.
Villiet D'Arignon (1568)

The flight left Miami after a 2-h delay. Soon after takeoff the turquoise line of the Florida Keys could be seen on the right, then the dark blue of the deep sea in the Straits of Florida—the narrowest point between Cuba and Key West—a treacherous channel crossed by thousands of Cubans in precarious rafts, the *balseros*,[1] seeking to escape from the miserable conditions on their beautiful island. Then, abruptly, came the announcement of landing at the José Martí International Airport in Havana. We could see the coastline with the surf approaching in rapid succession, followed by the light-green sugar fields, the patches of palm trees, and then the simple constructions of the airport in the old farm of "Rancho Boyeros." The plane came to a reverberating stop on the runway, and when the door opened we inhaled deeply the warm breeze filled with the salty scents of the sea.

An ancient red bus marked *Protocolo* met us on the tarmac disgorging a welcome party sent to greet the members of the Mission. We boarded the bus and were taken directly to a VIP lounge, where TV cameras, microphones, and flashing lights were waiting for us. The arrival of the Mission was receiving special attention from the national and international press.[2]

Cuba's Health Minister Dr Julio Teja Pérez and the PAHO/WHO representative in Cuba Dr Miguel Márquez greeted us. At the request of Minister Teja, Dr Llanos introduced the PAHO Mission team, outlined the objectives of the visit, and thanked the hosts for the welcome.[3] Next, Dr Jim Martone, on behalf of ORBIS, introduced his team and said to the journalists, "We're going to examine some patients and narrow down a list of possible causes and factors."[4] Dr Sadun, in answer to a journalist's request to describe the unique features of the Cuban epidemic, recalled previous epidemics of eye problems in Japan and Jamaica.

Soon we were spirited away, aboard the two light-blue minibuses belonging to the WHO. The wide avenue leading from Rancho Boyeros to Havana was almost empty of traffic that quiet Sunday afternoon. A few

Cuban Blindness. DOI: http://dx.doi.org/10.1016/B978-0-12-804083-6.00010-2

motorcycles and many black-colored bicycles shared the roadway. Havana, at first glance, looked like any sleepy tropical city.

Accommodations at the Biocaribe Hotel proved adequate. After a quick lunch, the Mission and the Cuban health authorities gathered in a conference room, and encountered their first point of contention. The Cubans intended to give the international visitors a tour of their showcase institutions. However, as agreed earlier that same morning in Miami, the PAHO and ORBIS visiting medical experts insisted upon seeing patients together, in particular those that were considered typical cases of the disease, in order to maximize the combined expertise of the two groups.

After a few tense moments, I expressed the opinion of my colleagues: "With all due respect," I told Vice Minister Dr Jorge Antelo, "We are not interested in visiting the new pharmaceutical factory."

We declared our immediate need to examine the patients in the hospitals and, more importantly, to visit them in their homes in the areas most affected by the epidemic—Pinar del Rio, first and foremost. This, the team explained, was crucial to gain an understanding of the disease and its possible causes. The tense moment that followed was defused by a humorous statement by the small, bespectacled Dr Rafael Muci-Mendoza:

"Mr. Minister," he said, "in this problem of vision, we have to go to the eye of the storm."

First, however, it appeared we *must* see at least one Cuban installation.

Center for Genetic Engineering and Biotechnology: Cubanacán, Havana

The modern towers of the Center for Molecular Biology and Biotechnology[5]—simply called, in Cuba, *Biotecnología*—are located in Cubanacán, a sparsely populated area of western Havana, close to several scientific institutes, hospitals, and the Faculty of Medicine. Green lawns and tropical gardens surround the fenced-in towers. Biotecnología represents the most intensive effort on the part of the Cuban government to gain entry into the promising new field of biotechnology and genetic engineering and has been called "the most technologically sophisticated research facility in the Third World." Biotecnología has been given top priority as one of Cuba's main strategies for economic survival.[6]

The first official activity of the Mission to Cuba was a briefing held at the Center. Biotecnología's directors greeted us at the door of the ample hall of the first building. A series of introductions followed, and we met a number of young people proficient in molecular biology, immunology, virology, physics, and related sciences. We learned that Cuba had prepared 20,000 biotechnology professionals who grew up under the Revolution and received free education and training in a world where the value of money was unknown and dollars were irrelevant. They told us that most of them come to work every morning pedaling on a heavy Chinese bicycle for 1 h in the humid heat of the tropical morning. They then take a shower at the Biotecnología and go to work with the vigor of monks dedicated to their chores, chants, and prayers, truly believing in the words of Colombian Nobel Prize writer García Márquez: "The greatest achievement of the human being is the proper formation of conscience, and that moral incentives, rather than material ones, are capable of changing the world and moving history forward."[7]

Nonetheless, during my time in Cuba, a significant number of health professionals we interacted with eventually requested refuge in Europe and the United States, either taking advantage of the travel opportunities provided by epidemiologic studies abroad or, less often, joining the masses of *balseros* risking their lives to leave behind a way of life without opportunities to achieve one's personal dreams.

We were then guided to an auditorium of ultramodern design, with comfortable chairs, simultaneous translation facilities, TV cameras, and projection equipment. A large audience, many in white laboratory coats, already filled the room. There followed a seemingly endless procession of presentations and introductions—members of the Civil Defense, officers of the Ministry of Health and the Cuban Academy of Sciences, ranking officials from the Politburo, and more.

Suddenly, soldiers in green fatigues, carrying automatic submachine guns burst into the auditorium and fanned out around the room. A whisper went through the audience, "Fidel!"

We were surprised by the unannounced arrival of Fidel Castro. The moment was electrifying. Here was one of the main living actors of contemporary world history in the second half of the twentieth century—a controversial figure both hated and loved, and a powerful, enigmatic leader who was a hero to some, villain to others, mystery to all. The Comandante was dressed in olive green uniform and black combat boots, his distinctive cap shading his Roman nose, the dark penetrating eyes, and his now gray-white beard. The epaulets on his broad shoulders bore the single white star of his rank, Comandante, in a field of two red-and-black triangles representing the July 26 Revolutionary Movement, and two gold olive branches. He carried no weapons in the wide dark-green military belt that closed his jacket.

The audience stood up in spontaneous applause as the leader came forward, walking slowly, his hands held together as if in prayer. He first approached Health Minister Dr Julio Teja.

"Did the delegations arrive without problem?" he asked.

"Yes, Comandante, they arrived after the usual delay."

"*¿Román, vino?* Did Román arrive?"

I was standing next to Minister Teja. The Comandante's question came as a major surprise. I was stunned to find the leader knew of my presence, or even my name.

Minister Teja made the introductions. Castro took my hand.

"Román, I have read your work," he said. "Very interesting articles on tropical neuropathies. The problems observed during the Spanish Civil War. The problems in Jamaica. They have helped us a great deal."

"Thank you, Comandante."

Apparently, at the outset of the current epidemic, Fidel Castro received the articles I had sent via Canada.[8] Obviously, he was informed of my presence among the members of the Mission.

Fidel Castro then moved along, greeting each one of the members of the PAHO/WHO Mission and the ORBIS group with a handshake and finally motioned for the Health Minister to join him in the podium to open the session. Minister Teja explained the organization of the briefing. Experts from different institutions would be presenting their findings about the epidemic, he said.

In the presentations that followed, members of the Mission and ORBIS were briefed on the history of the epidemic.

In November 1991, a number of men in their late 50s and early 60s—*guajiros* like José Polo Portilla—began losing their eyesight. They were all farmers from the tobacco-growing municipalities of Sandino, Güane, San Juan y Martínez, San Luis, and Consolación del Sur, all of which are located in Pinar del Río, the westernmost province of Cuba. In early 1992, the first eight of these patients sought medical attention at the ophthalmology service of the Abel Santamaría Hospital in the city of Pinar del Río.

Those eight cases were just the beginning. Attending physicians at that hospital—Drs Blanca Emilia Elliot, Marta María de la Portilla, and Carlos Perea—came across 12–36 new cases each month in the first half of the year. By July, the total number had climbed to 168 cases. Strangely, the doctors found that for every four men blinded by the disease, only one woman was affected and that practically all the patients were smokers and rum drinkers.

This large number of cases plus the clinical presentation—painless loss of vision in both eyes—was extremely unusual. Ophthalmologists in this part of Cuba usually came across one or two cases of such blindness in an entire year. The current onslaught of patients led them to report the cluster to the National Directorate for Epidemiology of MINSAP. Years earlier, during the last epidemic of meningococcal meningitis,[9] Cuba had organized a system of epidemiologic surveillance to respond immediately to reports of unusual numbers of cases of a disease. The response was swift, and the epidemiologists dispatched from Havana confirmed the cluster of blindness cases in Pinar and their characteristics: middle-aged men, tobacco farmers, smokers, and rum drinkers. The MINSAP epidemiologists created a local database of these cases, recommended consultation with experts in ophthalmology and neurology, and returned to the capital.

Shortly thereafter, a group of Cuban medical experts traveled to Pinar del Río to study the problem. The group included Dr Jaime Alimany, Professor of Ophthalmology; Dr Santiago Luis, Director of the National Institute of Neurology and Neurosurgery; and psychologist Rafael Moya from MINSAP. These experts concluded that the clinical presentation was inconsistent with an inflammatory disease of the optic nerve. They correctly pointed out that the overall process was more likely toxic or metabolic in nature—perhaps alcohol-tobacco amblyopia. Therefore, the experts called the disease optic neuro*pathy* (from the Greek word *pathos*, suffering) and recommended treatment with B-group vitamins, instead of antiinflammatory medications for optic neur*itis* (from the Latin *itis*, inflammation), which doctors in Pinar del Río had been using so far.

Despite the positive response observed in most early cases with the B-vitamin treatment, MINSAP was not ready to accept malnutrition as the cause of the outbreak, mainly because it could be seen as a criticism of the government and its incapacity to properly feed its people. Therefore, the epidemiologists decided, judiciously, to search for the presence of a possible toxic product associated with tobacco. Ninety percent of the first patients, after all, were cigar smokers. In July and August 1992, in cooperation with the laboratories of the Cuban Tobacco Institute and the National Center for Toxicology (CENATOX), comprehensive tests were performed to detect the presence of possible pesticide or herbicide residues in tobacco leaves or of foreign substances in cigars. Samples were taken for tests from the leaves of different *vegas* in Pinar del Río, and random cigar samples of different manufacturers were studied, as were cigarettes called "Popular," a brand which was imported from Italy and not made of Cuban tobacco at all but were widely consumed in the country.

Nicotine, tar, and other products of combustion were found to be within the expected range. There were no residual organophosphate pesticides, and the levels of carbamates and other insecticides were within the internationally allowed range. The MINSAP's epidemiology group concluded that the search for toxins in tobacco showed completely negative results.

But the number of cases continued to swell. The towns of San Juan y Martínez, San Luis and Pinar del Río were now reporting more than 90% of the cases. To complicate matters, Dr Carmen Serrano, Chief of Internal Medicine in Pinar del Río, began noticing other problems associated with the Cuban blindness.[10] Many of her patients also suffered from malaise, irritability, sleep problems, memory loss, depression, tongue and mouth ulcers, painful burning of feet and fingertips, and difficulty walking and urinating. These symptoms all suggested that in addition to the optic nerve damage, other nerves and the spinal cord were also involved.

Late that summer, in August, an epidemic outbreak of beriberi—a classic disease of malnutrition—occurred among the inmates of the Ariza prison; most were treated successfully with administration of vitamin B_1.[11] But elsewhere on the island the spread of the illness continued. It reached the cities, and the number of cases began escalating to truly epidemic proportions. In December 1992, a year after the first patients were seen by doctors, there were about 500 known cases. Three months later, in March 1993, cases were being reported from 6 of the 14 provinces of Cuba: Pinar del Río, La Habana, Cienfuegos, Holguín, Sancti Spiritus, and Santiago de Cuba. The total number of cases had risen to 4561.

It was at this time, on March 20, 1993, that Comandante Fidel Castro was finally informed of the epidemic of blindness ravaging the island.[12] Castro's outbursts of fury are legendary. When he learned that almost 5000 Cubans had gone blind and that he had been kept uninformed of the spread of the disease for over a year, he exploded with the rudest invectives. Dr Hector Terry Molinert, Vice Minister of Health for Epidemiology, was fired on the spot.

Dr Molinert was not some medical bureaucrat. In the 1970s, he had successfully organized the epidemiologic surveillance network that proved to be extremely valuable in the fight against meningococcal meningitis type B. Following numerous other successes, he was appointed Cuba's Director of the National Institute of Hygiene, Epidemiology, and Microbiology and later became Vice Minister of Health. As a direct result of his efforts, Cuba could boast one of the most efficient epidemiologic networks anywhere in the Western hemisphere.

On Castro's orders, however, Molinert was out. The directors of other institutes involved in the initial study of the epidemic were also fired summarily. It was then that Castro decided to confide in PAHO's Director Dr Guerra de Macedo, who advised him to inform the WHO fully and openly.

A short 4 weeks later—when Dr Antelo addressed the WHO assembly—the total number of cases had reached almost 26,000, with the numbers continuing to climb. Castro then ordered the creation of a national task force to be composed of members of every medical organization in the country that had anything to contribute.

Castro intended to beat the epidemic by approaching it as a military operation and not as a medical problem. Therefore, General Guillermo Rodríguez Del Pozo, Director of Civil Defense for Disaster Relief, was appointed chairman of the task force. A distinguished military physician trained in the Soviet Union, he had been in charge of sanitary services for Cuban personnel in Angola and other African countries. General Rodríguez was used to making rapid decisions, setting priorities, and getting the job done quickly and effectively. There would be daily reports from the different components of the task force in meetings to be conducted in a military situation room. Specific tasks would be given to each of the participating components, and deadlines would be strictly enforced. Even the name of the task force, *Grupo Operativo Nacional*, National Operative Group—where the word *neuropathy* was absent—indicated the military nature of the group.[13]

The Minister of Health and the President of the Cuban Academy of Sciences were ordered to coordinate the 55 participating institutions grouped into six research areas, including epidemiology, toxicology, basic sciences, nutrition, clinical and therapeutics, and food resources. The main hurdle faced by the previous epidemiology group studying this epidemic had been the lack of priority for their activities at a time when the island was dealing with problems related to the *período especial*. This had now been corrected by the Comandante's direct intervention. Activities related to solving the epidemic became Cuba's top priority.

The Mission's briefing session ended on a sobering note.

"Despite our best efforts, the epidemic curve continues to climb," Minister Teja said. "We believe this is a new disease. We have been unable to find previous references to epidemics of such magnitude and with these clinical characteristics. By helping Cuba today, you are helping humankind."

Fidel Castro remained silent, deep in thought.

Date: Sunday, 5/16/1993 Hotel Biocaribe, Havana, Cuba: It took me a long time to recover from the surprise meeting with Fidel Castro after only a few hours in Cuba. It was totally unexpected that Castro asked if I had arrived with the Mission and even more when he told me that he had enjoyed reading some of my old papers on tropical neurology where I had shown their similarity to diseases described by Spanish physicians during the Civil War in Spain. Castro has remained interested in his country of origin. His father was born in a farming village in Galicia, Spain; in 1898 he came as a soldier of the Spanish Army to combat Cuban independence and decided to stay on the island after the war. Fidel had visited his family's old home in Galicia a year earlier, in July of 1992.[14]

The events of that night impressed upon me the gravity of this nation's epidemic, of Castro's approach to dealing with it, and of the Mission to Cuba.

★★★

As part of the Mission's briefing, Dr Gustavo Kourí—son of the famous parasitologist Pedro Kourí, founder of the Institute of Tropical Medicine in Havana—had explained in his powerful baritone voice that a number of samples of blood and cerebrospinal fluid collected from patients with Cuban epidemic neuropathy had been studied for the presence of a microorganism—a virus, bacteria, fungus, or parasite—that could be causing the outbreak. One of their laboratories had isolated an unknown viral agent, but it was clear that the Cuban health professionals felt insecure about proposing it as the cause of the epidemic. The main reason to rule out a transmissible agent was the absence of a pattern of contagion, which is typical of infectious epidemics, be it cholera or the flu. In such epidemics, one case infects close contacts, and the contagious disease spreads like fire in a dry forest. None of this was happening in Cuba, where all cases were isolated and without contact among the patients. Admiral Young could be reassured that quarantine was not required. But this was also the reason for the Cubans' request for the presence of a virologist as part of the Mission to Cuba team.

Dr Kourí's point was well taken. As the briefing ended, Dr Llanos and I discussed the urgent need to find a virologist with experience in pathogenic agents that affect the nervous system. I proposed the name of Dr Carleton Gajdusek. He was my colleague at the National Institutes of Health and (NIH) had received the Nobel Prize in 1976 for his discovery of prions, a new type of viruses that destroy the brain, and he was one of the world's leading neurovirology figures.

The next day, Dr Gajdusek was presented with our request and he agreed to participate. He and two other NIH colleagues, Drs Paul Brown and David Asher, would join us in Havana in 2 days.

ENDNOTES

1. *Balseros* = rafters (from the Spanish *balsa*, rafter); name given to the persons who emigrated illegally in self-constructed or precarious vessels from Cuba to neighboring states (Wikipedia). Catalan film, Academy Award nominee for Best Documentary Feature (2002), directed by Carles Bosch and Josep Maria Domènech, written by Carles Bosch and David Trueba (Wikipedia). Holly Ackerman. *The Cuban rafter phenomenon. Selected bibliography.* Duke University, Latin American & Caribbean Studies. www.guides.library.duke.edu and holly.ackerman@duke.edu.
2. Larry K. Altman. 26,000 Cubans partially blinded; cause is unclear. *The New York Times* (May 21, 1993); US doctors in Cuba to study puzzling nerve ailment, Reuters (May 17, 1993).
3. "Venimos en misión de explorar y trabajar conjuntamente," expresa representante de la Organización Panamericana de la Salud (OPS) al arrivar al frente de un grupo de especialistas que colaborarán en el enfrentamiento de la neuropatía epidémica. *Trabajadores* (La Habana, May 17, 1993); JA Martín. Expertos de OMS/OPS y Proyecto Orbis recorren centros hospitalarios. Constatan manifestaciones clínicas de la neuropatía epidémica. Sostienen intercambios científicos con personal médico cubano, *Granma* (May 18, 1993).
4. Quoted in, US doctors in Cuba to study puzzling nerve ailment, *Reuters* (May 17, 1993).
5. CIBG. Historia del Centro de Ingeniería Genética y Biotecnología, Apartado 6162, Cubanacán, La Habana, Cuba.
6. Julie Feinsilver. Can biotechnology save the revolution? *NACLA Report on the Americas* 26(5):7–10 (May, 1993). SmithKline Beecham obtained the production rights of the new Cuban anti-meningococcal vaccine; see, Adams D. Vaccine tests Cuba policy. *St. Petersburg Times* (May 9, 1998).
7. Gabriel García Márquez. An encounter with Fidel. In, *Inside guides: Cuba*, pp 52–53 (Singapore, 1995).
8. Castro was referring to the following articles: Gustavo Román, Bruce S. Schoenberg, Peter S. Spencer. Tropical myeloneuropathies: The hidden endemias. *Neurology* 35(8):1158–1170 (1985), and to a condensed version in Spanish published as, Gustavo Román. Mielopatías y mieloneuropatías tropicales. *Boletín de la Oficina Sanitaria Panamericana* 101(5):452–464 (Noviembre, 1986).
9. Valcarcel Novo M, Rodríguez Cruz R, Terry Molinert H. *La Enfermedad Meningocócica en Cuba. Cronología de una Epidemia* (Ciudad de La Habana, Cuba, 1991).
10. Carmen Serrano, et al. *Informe del primer año de neuropatía epidémica en la provincia de Pinar del Río.* Typewritten, pp 68 (December, 1992).

11. Alfredo Espinosa, et al. (Op. cit.).
12. According to Minister Teja, "On 20 March 1993, our Commander-in-Chief, who had not been duly alerted, called an urgent meeting to analyze the problem, and to organize, direct, and take each and every one of the steps required to curb the epidemic, and to invest all efforts, resources, and technical and scientific know-how in this fight against the epidemic, and to inform and advise the population." (Julio Teja. *Closing Remarks, International Workshop on Epidemic Neuropathy* (July 15, 1994). In, F. Rojas Ochoa (ed.), *Neuropatía Epidémica en Cuba* (La Habana, Cuba, 1995, p 229).
13. Grupo Operativo Nacional. *Neuropatia Epidemica en Cuba.* La Habana, Cuba, 30 julio 1993; Esequiel Hernández Almeida. La epidemia de neuropatía Cubana: un fenomeno reemergente. *MULTIMED* 1998;2(3).
14. Pablo Alfonso. La casa matriz de los Castro está en ruinas. *Martí Noticias* (Oct. 31, 2012): En julio de 1992 Fidel Castro viajó a Láncara para conocer la casa de su padre, Ángel Castro, aprovechando la visita a Galicia invitado por el entonces presidente de la Xunta y lider del Partido Popular Manuel Fraga. See also, Tom Harvey. Visit to Spain shows isolation of Fidel Castro. *Sun Sentinel* (July 30, 1992).

CHAPTER 11

Epidemiologists: Medical Detectives

Elementary, my dear Watson.
Sherlock Holmes

With the briefing behind us, Mission members and ORBIS experts alike agreed that it was impossible to determine the accuracy of Minister Teja's conclusion that the epidemic of blindness was a novel disease being caused by a new viral agent. Based on the clinical descriptions presented at the briefing, the identity of the disease was still blurred and would remain so until we were given the opportunity to examine the patients. We would have to talk with the patients and their families, visit their homes and places of work, and, most importantly, examine as many of them as possible to find the cause of the epidemic.

Date: Monday, 5/17/1993, 01:00 hours, Hotel Biocaribe, Havana, Cuba: Hotel Biocaribe, Havana, Cuba: Tonight, here at our hotel, over dinner, we had an animated discussion on the events of the day and our first impressions. The presence of Castro at the briefing and the organization of the task force indicate that the epidemic is Cuba's number one priority. Also, the quality of the epidemiologic information is quite good. We also learned that we would be meeting nightly with the Cuban Neuropathy Task Force at the hotel's large conference room on the second floor.

Epidemiologists at work use a pattern of reasoning that is quite similar to that of detectives solving a murder case. When the police arrive at the scene of a crime, they must first ask, "Who is the victim?" All the characteristics of the victim—even those that later on may be found to be irrelevant—are carefully noted: age, gender, race, occupation, social status, education, travel, family history, and so on.

Next, the detectives examine the scene of the crime, looking for signs of violent entry, possible weapons, and the presence of blood or fingerprints. In medicine, the equivalent facts are the circumstances surrounding the beginning of the disease. The numerous questions the doctor asks the

Cuban Blindness. DOI: http://dx.doi.org/10.1016/B978-0-12-804083-6.00011-4

patient are directed toward the search for clues indicating the nature of the disease. Was there fever, indicating infection? Was there weight loss or fatigue, pointing to a malignancy?

In a murder case, the forensic pathologist arrives last to examine the body and provides essential data: time and cause of death, number and type of wounds, evidence of torture, sexual violence, drugs, toxins, and presence and type of blood, semen, hair, and so on. Likewise, the physician's examination of the patient is an active search for diagnostic clues.

Detectives begin to identify suspects from the onset of their investigation. In the same way, a good clinician begins to sort out a number of tentative diagnostic hypotheses while first examining a patient. The examination actually is a search for signs to confirm or deny the successive hypotheses in order of importance, from the most likely ones to the really far-fetched.

But finding the actual culprit is a complex affair. Files on thousands of possible suspects are reviewed to match the *modus operandi* of a number of felons whose crimes had the characteristics of the present one. Finally, after getting all the relevant information together, the suspects are listed and their alibis verified.

Epidemiologists are medical detectives dealing with epidemics of diseases.[1] They keep track of diseases occurring in the community and describe their patterns of injury and attack rates, and they learn to recognize the peculiarities of each one of the diseases—its preference for occurring in summer or winter, for example, or for attacking children or older adults. Some disease agents are gender-fixated villains, attacking only males or only females. Others are rogues obsessed with social class differences, such as the agent of tuberculosis *Mycobacterium tuberculosis*, which prefers indigent homeless shelters, or *Giardia lamblia*, a parasite that feeds upon guests in exclusive ski resorts. A few vicious attackers cause fulminating death, while others are slow, persistent agents causing chronic disease. These features are called the "natural history of a disease."

For the epidemiologist, the natural history of diseases is the medical police file where the criminal records of each disease are carefully kept, with their particular fingerprints and mug shots. But, even with computer files, the true epidemiologist—like a good detective—has to walk the streets to find the culprit.

The following morning, after a quick breakfast of coffee, pastries, and tropical fruits, the Mission team went to the dilapidated MINSAP building in downtown Havana to meet with the Cuban epidemiology group. We were then presented with a detailed, computer-driven report—illustrated with state-of-the-art color graphics—of the numerous studies conducted to date on the island from onset of the epidemic.

The studies had been carefully planned and executed. The visiting experts found no methodologic flaws. The statistical tests used for data analyses also appeared to be correct. Dr Llanos requested and obtained additional computer time to review some specific points of analysis with the statisticians. The database on the epidemic was huge and included information from every municipality on the island. Data were updated daily via telephone. The total number of patients with confirmed epidemic neuropathy had reached 33,966 cases, on this date, with more than 4000 new patients every week.

Contrary to our initial fears that epidemiology in Cuba would be rudimentary at best, we found that the surveillance network had been organized in a very professional manner, and the flawless handling of the information spoke eloquently of the high level of development of epidemiology in Cuba. Mission members, who, from experience in other developing nations, were prepared to begin with the ABC of epidemiologic cluster analysis, were delighted to be able to discuss the data at an advanced level.

The most striking finding of the epidemic continued to be the predilection of the for those in the middle of life, with no victims among children, teenagers, pregnant women, and older adults, groups that are usually the first victims of infection, famine, and hunger. Also, team members immediately made another interesting observation: There was a persistently low prevalence of the disease in the Guantánamo Province, in particular in the municipalities of Caimanera, Maisí, El Salvador, and Baracoa, as well as in the *Isla de la Juventud*, island of youth (Pine Island). This finding suggested that either these populations were somehow protected or the action of the causative agent was less aggressive in those areas. I remembered the words of the Cuban diplomat Pablo R. Rodríguez, who had contacted me in Bethesda and hinted that foul play was suspected. The geographic distribution of the disease probably had been construed as circumstantial evidence that GITMO had been spared. Therefore, the Mission requested a visit to Isla de la Juventud—the least affected area—following the trip to Pinar del Río, which continued to be the area most affected by neuropathy.

Also of particular interest was the very steep upturn in the epidemic in March 1993, around epidemiologic week 12, that Mission members noted. But what at first seemed a potentially important clue proved, instead, to be of only passing interest. Since early cases of the neuropathy had shown the best response to vitamin treatment, the *Grupo Operativo* had decided to use Cuba's family doctor program for the active search of patients who were just beginning to have symptoms of the disease. In other words, starting in March 1993, with directives from the *Grupo Operativo*, all family doctors began actively looking for, and finding, early cases of Cuban blindness.

Cuban family doctors living in the community are Cuba's first line of the early warning system of the island's epidemiologic network and their response was overwhelming. In 2 weeks, from March 28 to April 10, an increase of 48% in reported cases was confirmed. All reported cases were examined by specialists to confirm the diagnosis, and those with positive diagnoses were admitted to the hospital for treatment. The case-finding measures strained Cuba's limited reserves of finances and supplies due to the *período especial.* According to Vice Minister Abelardo Ramírez, "the expenditures were more than US$100 million dollars, represented in equipment, urgent opening of 20,000 new hospital beds—an increase of 30% of all available beds, often having to be placed in schools and community centers—along with the emergency purchase and airlifting of vitamins, syringes, needles, and supplies."[2]

At the end of the second day we had learned all that we could from the classroom. It was now the time to go into the field and meet face to face with the victims of the epidemic.

ENDNOTES

1. Berton Roueché. *The Medical Detectives* (New York, 1991); P. Jaret. The Disease Detectives. *National Geographic* 1991; 179 (1):114–140.
2. Abelardo Ramírez Márquez, Pedro Más Bermejo, G. Mesa Ridel, J. Hadad Hadad, A. Marrero Figueroa, E Zacca Peña. Síntesis de los principales aspectos y manejo de la neuropatía epidémica, In: F. Rojas Ochoa, editor. *Neuropatía Epidémica en Cuba* (La Habana, Cuba, 1995); Grupo Operativo Nacional. *Neuropatia Epidemica en Cuba.* (La Habana, Cuba, 30 julio 1993).

CHAPTER 12

Eye Hospital of Havana

And out of the gloom and darkness the eyes of the blind will see.
Isaiah 29,18

Early next morning the Mission members and their ORBIS counterparts were whisked off from their hotel to Havana's huge Ramón Pando Ferrer Eye Hospital. The day was already hot and humid. Outside the front door of the hospital, long lines of men in white *guayabera*[1] shirts and women in light-colored dresses anxiously awaited their turn to be examined. A few were already blind, eyes open but not seeing, heads bowed in a timid apology, and listening with one hand cupped around the ear and the other resting on the shoulder of the guide.[2]

After some preliminary formalities and the usual offering of strong-and-sweet *café Cubano*, the expert group was conducted to a classroom. Dr Carolina Salazar, a retina specialist who had been appointed director of the hastily organized *Servicio de Neuritis Optica* (Optic Neuritis Service), made a presentation of the known clinical data. In the past few weeks she had personally examined 861 patients with epidemic optic neuropathy.

Most patients complained of loss of eyesight or blurred vision. None complained of eye pain, but all found the bright sunlight of the tropics too intense, almost painfully so, forcing them to stay indoors, and when outdoors to wear hats and caps and to use the dark glasses favored by the blind.

Dr Salazar had carefully examined all sorts of patients, from those who had noticed loss of vision just a few days earlier to others who were almost blind after 8 months or more of progressive dimness of vision.

Her research revealed three consistent problems: (i) loss of visual acuity,[3] (ii) loss of color vision for red and green,[4] and (iii) loss of the central field of vision, which had been replaced by a circular blind spot.[5] This explained the common complaint that patients could no longer recognize a familiar person until they could identify the person's voice.

There were other symptoms, not shared by all of the patients, which suggested to me that something else was amiss, that the disease was attacking them beyond their optic nerves. About one-third of patients were becoming

Cuban Blindness. DOI: http://dx.doi.org/10.1016/B978-0-12-804083-6.00012-6

prematurely hard of hearing, and most of them had noticed an unpleasant pins-and-needles tingling sensation in their hands and feet.

Dr Salazar's final words confirmed my suspicions.

"All these patients, all of them, have behavioral changes," she said. "They have become irritable, restless, and constantly tired but unable to sleep, and a large majority of them have been prescribed meprobamate, diazepam, phenobarbital, or similar medications for weeks or months to control these symptoms."

With the introductory speech concluded, we were now allowed to see patients. Dr Salazar had selected a complete gamut of cases—ranging from some with mild early disease to some with advanced severe cases; men and women; young, middle-aged, and older patients; and smokers and non-smokers. Two or three members of the mission interviewed every patient.

Dr Caballero concentrated on dietary data. Dr Muci-Mendoza investigated past medical problems and other manifestations. I conducted a standard neurologic examination. Drs Sadun, Martone, and Silva examined the eyes. In a separate room, neuroophthalmology technician Lillian Reyes had set up Dr Sadun's portable equipment to photograph the retina and to conduct specialized tests.

Professor Sadun began examining the concave surface—the *optic fundus*—deep inside the patients' eyes by using the ophthalmoscope, an instrument that visualizes the retina, the retinal arteries and the veins, the cream-colored circle of the optic nerve, and also the axons or nerve fibers converging toward the optic nerve from all corners of the retina.

Professor Sadun held the ophthalmoscope in one hand, approached the first patient, took a quick glance at both eyes, moved rapidly to the next patient, and then to the next.

Excited, and smiling broadly, he announced, "I knew it! All these patients have a typical lesion in the optic fundus. There is a wedge-shaped defect produced by the loss of the maculopapillary bundle."[6]

The presence of this retinal defect was rapidly confirmed by all the Mission members and by the Cuban ophthalmologists, who began calling this wedge-shaped retinal defect "*La cuña de Sadun*" or "Sadun's wedge sign."[7] It provided the pathognomonic element, the *sine qua non* of the Cuban optic neuropathy, which Dr Jim Martone had been searching for.[8]

Professor Sadun recommended the use of polarized glasses, which he had invented for the diagnosis of lesions similar to those he had just found in Cuba.[9] These measures allowed the Cuban physicians to rapidly confirm the presence of the disease in a given patient.[10] Absence of the wedge sign would exclude from the epidemic totals those patients suffering from other causes of loss of eyesight such as macular degeneration and genetic diseases such as Leber hereditary optic neuropathy. It would exclude, too, any cases of malingering or hysterical blindness, in subjects who wanted to improve the diet during a few days in the hospital.

That afternoon, following our successful visit to the Ramón Pando Ferrer hospital, we were taken by bus to the neighborhood known as Siboney.[11]

Prior to the Revolution, it had been the most elegant suburb of Havana. Since the epidemic, though, a number of its modern houses had been transformed into dwellings for patients with neuropathy.

Nearby was the CIMEQ,[12] the Medical and Surgical Research Center hospital, considered the top clinical research facility in Cuba and an exclusive destination for "medical tourism" for patients from all over the world.

A permanent military detachment guarded the entrance. Inside were diplomats, members of the government and other important persons, and medical tourists from many nations. Transplantation of fetal brain cells to treat Parkinson disease produced a constant influx of patients from Argentina, Spain, and Italy. Children with cerebral palsy underwent rehabilitation treatment, made famous by the physical therapy techniques that had produced Cuban Olympic champions such as the undefeated super heavy weight boxing champion Teófilo Stevenson, Gold medalist in the 1972, 1976, and 1980 Olympics.

Medical tourism was a thriving new high-tech industry, providing a source of much-needed foreign currency. Outside, its white walls and red tile roofs gave the Siboney neighborhood the look of a tropical country club. Inside, the business was clearly medical care. With the onset of the blindness epidemic, the most advanced clinical research techniques were being utilized for the evaluation and study of patients with neuropathy.

Patients were plentiful in the elegant neighborhood of Siboney. We went from house to house examining them all. These patients suffered from a disease that was identical from one case to the next, and the next, over and over again. For the researchers, it was a strange experience. We were seeing multiple cases of a disease so rare that, until that moment, the combined experiences of the entire team had amounted to no more than one or two cases in a lifetime of medical practice.

However, the surfeit of optic neuropathy patients and the enormous amount of data on their disease did not immediately provide a clear solution to the problem. The CIMEQ researchers, presenting their findings

around a large conference table, over more cups of *café Cubano*, had failed
to produce a single clear indication of the cause of the epidemic.

There was no consistent trend in the changes observed in the patients'
blood counts, bone marrow examinations, blood chemistries, liver and
kidney function tests, immune function tests, cerebrospinal fluid examinations, or measurement of nutritional values. Low values of vitamin B_1 or
thiamine had been found persistently in patients, but these were also present in the "healthy" population.

The patients themselves presented a grim picture. How old they
looked! Men and women in their 40s seemed prematurely aged, looking
more like someone in their late 60s. They all seemed to have lackluster
eyes; protruding bones under their sunburned, parchment-thin skins; thinning gray hair; sad, toothless mouths; nicotine-stained fingers; and a look of
dejection. A few smelled so strongly of rum that alcoholism was a strong
possibility.

PEDRO JAVIER, PELOTERO

I was particularly impressed by one of the patients living in Siboney. Pedro
Javier was a 45-year-old man, who, during his high-school days, had
been an excellent *pelotero*,[13] an accomplished baseball pitcher, who participated in the Pan-American Games with the Cuban team. Pedro Javier
had worked with one of the local Committees for the Defense of the
Revolution (CDRs). His room displayed a trophy and a medal collection.
His most valuable souvenir was a Red Sox batting practice hat that he
had gotten from a scout, who offered to bring him to Boston immediately after the Pan-American Games. He had refused the offer at once and
without hesitation, but in his mind the *B*-hat became his symbol of vali-
dation. For many years thereafter, he had served as assistant coach of the
Cuban National team.

I asked him, "Pedro Javier, tell me, how did your problems begin?"

"Well, doctor, it began with stomach cramps and diarrhea 5 or 6
months ago. You know, in this island having diarrhea is nothing strange.
There are times when the *camaradas* at the Aqueduct have no alum to
purify the water, and then everybody in town get the runs for a few days.
I continued to have diarrhea several times a day for almost 1 week and
went to see my local family doctor. She prescribed some kaolin and paregoric, and the problem was controlled for a couple of days. I stopped the

medication, and sure enough, the loose stools returned, twice or sometimes three times in the day with cramping pain in my belly. More kaolin
and paregoric did the trick again for a few days, but the diarrhea never
really stopped. One night, maybe 3 months ago, I began to feel constant
tingling in my fingertips and in the soles of my feet, and I noticed that
I had lost the sense of touch in my hands; it feels like if I am wearing
white gloves all the time. After so much time with diarrhea I lost almost
30 kilos, lost all my energy, and had no appetite. The worse part was the
burning pain in my feet; it was like your feet were scalded by hot steam,
vapores giving you the feeling that the skin has peeled off. The raw feet
feeling was there all the time; it was worse with walking, so I stayed in the
recliner or in bed all day long. The pain was worse at night, and I would
just sit in the dark, grabbing my feet to ease the pain, unable to sleep and
on the verge of tears. Of course, I was exhausted in the mornings from the
pain and the lack of sleep and also because after I had finally fallen asleep,
I had to get up to pee several times. During the day I was constantly dizzy
and noticed that I was losing my hearing, my hair, and my teeth. When I
began to have problems reading, I knew that I had developed epidemic
neuritis and went to my local hospital, and they sent me to the CIMEQ
for evaluation. They kept me in the hospital for 10 days, gave me vitamins
in the vein, and then sent me to this house. I feel a little better now."

"Tell me, Pedro Javier, do you smoke?" I asked.

"Well, doctor, I really never smoked or drank alcohol, since I was
into sports all my life; but with the embargo, like many others, I began to
smoke to ease the hunger cravings. But I never smoked more than one or
two cigarettes a day; a pack of *Popular* lasts me almost 1 month."

"Pedro Javier, let me examine you," I requested.

The powerful body of the former pitcher had suffered severe assault
by hunger. The round baby face seen in the photos on the walls had been
replaced by sunken cheeks and prominent cheekbones; the eyes were
sunken and sad and surrounded by dark shadows; his hair was gray and
scant, his skin was furrowed, and he looked many years older than his 45
years, which he was about to celebrate. He still had the physical frame and
the broad shoulders of his youth, but the muscles had melted away, and I
was able to easily break the resistance of his pectoralis, biceps, and deltoids.
The legs were also weak, and his feet still felt raw. He had no reflexes in
his arms or legs, and he had lost all sensation in his hands up to his mid-
arms and below his knees in a pattern that neurologists call "stockings and

gloves." I examined his eyes with the ophthalmoscope and saw Sadun's wedge and minimal pallor on the external side of his optic nerves. With treatment, his vision problems had been reduced to some residual difficulty reading the Ishihara color plates, suggesting that the cones in the *macula lutea* of his retina still had not recovered completely. Likewise, he had some problem hearing the high-pitch tuning fork.

"Pedro Javier, I think they treated you on time," I said. "In my opinion, you have peripheral neuropathy from lack of vitamins and from smoking; you will need to continue receiving the special high-protein diet they are giving you, and you need to take the multivitamin pills every day. No more *Populares* for you, and no beer or rum. Once you begin to regain your strength, you will have to exercise again, slowly. If the diarrhea returns, do not let it go untreated because you will lose the nutrients faster than you can replace them in the diet and you could have a relapse."

I got up to leave and said, "Pedro Javier, it was a pleasure meeting you, and I wish you the best of luck."

I offered him my hand.

"*Gracias Doctor Román,*" Pedro Javier replied with a sad smile.

★★★

What price, I wondered, are the Cuban people paying for their Revolution?

Night after night, an estimated 11 million Cubans were going to bed still hungry after eating their meager portion of *arroz y frijoles*, prisoners' food: *comida de preso*. Most were resorting to eating sugar by the fistful—the only inexpensive and abundant commodity on the island—to alleviate the hunger pains.

Despite their hunger, every single patient, in their private confessions during the medical interview and examination, truthfully mentioned the rare consumption of meat, coffee, dairy products, eggs, fish, pork, fruits, vegetables, or oil. The patient's answer was always vague and embarrassed, and preceded by unsolicited explanations and a defense of their nation's egalitarian society.

"You know, the situation is very difficult," the patient might say. "With the blockade, we are in the middle of the *período especial*. Nonetheless, every day, for everyone on the island there is *arroz y frijoles*. The other things we haven't had for months. But since the triumph of the Revolution, nobody dies of hunger in Cuba—even with the embargo."

NUMANTIA

When the legions of the Roman Empire began the conquest of the Iberian Peninsula, they found firm resistance from a group of native Celtiberians that inhabited the fortified city of Numantia, located on the banks of the upper Duero River in Spain.[14] Beginning in 147 BC, and for the next 60 years, three Numantine generations resisted the invaders and fought ferociously against the attacks of the Roman armies. The Numantines first defeated 30,000 Roman legionnaires under two different Consuls and then an army of 20,000 men. Year after year the fortress withstood the attacks and repelled the Roman troops.

"Numantia was a sort of Vietnam for Rome," writes Carlos Fuentes.[15] "Its lack of success demoralized the Roman army, the Roman public protested furiously at the prolongation of war, which devoured draft after draft of young men, and the Senate refused to send any more troops."

How could this people, these "barbarians" poorly armed and ill equipped, oppose for so long the otherwise invincible Roman legions, the most powerful army of its time?

Fuentes answers: "Discovering that their strength lay in defense, the Spanish refused to offer a visible front line and instead invented guerrilla warfare. Surprise attacks by very small bands, preferably at nighttime; armies that became invisible by day, blending into the whitewashed villages and the gray mountainsides; dispersion, counterattack—these made up *la guerrilla*, literally, the little war."

Finally, in 134 BC, the Roman Senate, ashamed by the incompetence of the army, offered the command to Scipio the Younger, the conqueror—and annihilator—of the rebellious city of Carthage on the north coast of Africa.[16] Elected Consul, he was sent to Spain with 60,000 men to crush the rebellious Numantines. Scipio was committed to winning the war at all costs. "From the moment of Scipio's arrival the army in Spain felt his iron hand," says Leonard A. Curchin, in *Roman Spain*.[17] Scipio built a wall, 3 m high, around the periphery of Numantia, with ramparts that closed the grip around the city. He then burned the surrounding fields as part of his plan to starve the Numantines into submission. The 6000 defenders of Numantia attempted in vain, once and again, to unlock the siege, but the superior Roman forces withstood the attacks. Finally, starvation defeated the proud Celtiberians. The historian Appian, in his *Roman History*,[18] recounts the final days of Numantia:

*The Numantines, being oppressed by hunger, sent five men to Scipio to ask whether he would treat them with moderation if they would surrender…. Scipio…said merely that he would only accept **deditio**, unconditional surrender… Soon after this, all their eatables being consumed, having neither flocks, nor grass, they began, as people are sometimes forced to do in war, to lick boiled hides.*

When these also failed they boiled and ate the bodies of human beings, first of those who had died a natural death, chopping them into small bits for cooking. Afterwards, being nauseated by the flesh of the sick, the stronger laid violent hands upon the weaker…. In this condition they surrendered themselves to Scipio.

In the winter of 133 BC, the Romans took the city. Many Numantines—like the Jewish Zealots at Masada resisting Roman rule in the first century AD—preferred suicide to surrender. Appian wrote:

The majority of the inhabitants having killed themselves, the rest…came out… offering a strange and horrible sight, their bodies dirty, squalid, and stinking, their nails long, their hair unkempt and their dresses repugnant. If they seemed worthy of pity because of their misery, they also inflicted horror because on their faces were written rage, and pain, and exhaustion.

Even in defeat, Numantia became the symbol of the fiercely independent Iberian spirit facing the conquering Roman Empire.[19]

The Cuban people, who were children and grandchildren of Spaniards, had knowledge of Numantia in their hearts.

"We shall follow the example of Numantia," Fidel Castro told the Cubans.[19] He constantly reminded them that, like the Numantines, they would never give up the fight: It was a question of *pundonor*.[20]

Date: Monday, 5/17/1991, 22:00 hours, Hotel Biocaribe, Havana, Cuba: Tonight, we had the first debriefing session in the conference room on the second floor of the Biocaribe Hotel. Comandante Castro and Minister Teja presided over the session. Professor Sadun was in the spotlight. His findings of the retinal wedge defect and the additional diagnostic accuracy of the polarized glasses were received with great interest. Instruction courses on the use of the lenses and on the characteristics of Sadun's wedge were to be immediately planned for all the ophthalmologists assigned to the epidemic. The best technical photographer in Cuba was given the task of reproducing the test materials to be used with polarized lenses.

Cuban ophthalmologists—who had seen the wedge defect but minimized its importance—wore the dejected look of a baseball team that had just lost a critical game by one career.

The Mission to Cuba concluded its second day on the island having made two decisive contributions to the epidemic: (i) True cases of the disease can now be identified with certainty; and (ii) the mechanism of disease begins to be unraveled with confirmation of the selective loss of cones in the retina. I am beginning to sense some progress, but the pace might be too slow for the thousands of Cubans who are becoming blind. Also, the disease appears to have no clinical or epidemiologic resemblance with any known textbook condition. And, to complicate the problem even further, we were advised tonight that Dr Pedro Mas Lago, the veteran virologist of the Kourí Tropical Medicine Institute, would presenting to us the next day some details concerning an enterovirus isolated from the cerebrospinal fluid of patients with Cuban epidemic neuropathy.

ENDNOTES

1. *Guayabera* = A lightweight open-necked Cuban or Mexican shirt (Oxford).
2. *Lazarillo* is the word commonly used in Spanish for the guide of a blind person. The name was popularized in a famous anonymously authored novel called *Lazarillo de Tormes* (1554) considered the most influential picaresque novel in Spain (Wikipedia).
3. *Visual acuity* is the technical name for the capacity of the eye to see images sharply, to "resolve" images. It is measured, by international convention, using the familiar Snellen test chart. Placed at a distance of 20 feet, or 6 m, the letters of diminishing size can normally be read at distances of 200 feet to 10 feet. Normal visual acuity is 20/20; one-third of Cuban patients had 20/200 vision and at 20 feet could see only the tallest letter "E" in the chart. Many patients were legally blind. Improper focusing by the optical elements in the eye is the most common cause of poor visual acuity, and this can be corrected with spectacles. Not in Cuban patients. Use of a variety of corrective lenses failed to improve the eyesight. This suggested that the damage was more serious, perhaps an injury to nerve elements in the macula, the center of the retina where images converge—the sensitive film inside the camera-like structure of the eye.
4. The retina is a light-sensitive membrane lining the eye. Embedded in the retina are light receptors called *rods* and *cones*. Rods are abundant (110–125 million), reacting to light of low intensity, peripheral vision, movement, and night vision. Throughout evolution their function has been vital for survival: hunting and defense. Cones are less numerous (6.3–6.8 million) and are concentrated in the macula. In contrast to rods, cones are stimulated by high-intensity light and are concerned with color vision and fine details. The loss of color vision for green and red in patients with Cuban blindness pointed to the loss of cones or their connections. Also, the cones-rich macula is the point of clearest vision explaining the loss of visual acuity.
5. The macula can be examined by using either the artistic color figures designed by the Japanese ophthalmologist Ishihara to detect familial color blindness or by using a black screen called a *perimeter*. With one eye covered and focusing on the center of the screen, a normal person has about 100° of lateral peripheral vision, 75° downward

and 50° toward the nose and upward. The normal visual field is then egg shaped and slightly tipped upward. Just off center there is a small blind spot that corresponds to the point of attachment of the optic nerve. Cuban patients failed both the Ishihara test and the perimetry examination. More surprisingly, the central point of vision had been replaced by a large blind spot of about 5° giving the visual field the appearance of a bull's eye target.

6. The maculopapillary bundle is a skein of nerve fibers or axons connecting the cones in the macula to the papilla or head of the optic nerve and thence to the brain.

7. Professor Alfredo Sadun received the Cuban National Medical Academy's Medal of Honor, awarded for his contributions to the study of the epidemic blindness in Cuba. See, Ocular Surgery News. Dr Sadun honored for work during blindness epidemic in Cuba (March 1, 2002).

8. Alfredo A. Sadun, Martone JF, Muci-Mendoza R, Reyes L, DuBois L, Silva JC, Román GC and Caballero E, Epidemic optic neuropathy in Cuba: Eye findings. *Archives of Ophthalmology* 1994; 112:691–699.

9. Wall M, Sadun AA, Threshold Amsler grid testing: Cross-polarizing lenses enhance yield. *Archives of Ophthalmology* 1986; 104:520–523.

10. Christina Mills. In the eye of the Cuban Epidemic Neuropathy storm: Rosaralis Santiesteban MD PhD. Chief of Neuro-ophthalmology, Neurology and Neurosurgery Institute. *MEDICC Review* 2011 (January) 13(1): 10–15.

11. The name "Siboney" was popularized in a 1929 song by Cuban composer Ernesto Lecuona; it was used in the score for Fellini's film "Amarcord" (Wikipedia).

12. CIMEQ = Centro de Investigaciones Médico-Quirúrgicas.

13. *Pelotero* = common name for a baseball player in Cuba and the Caribbean.

14. The ruins of Numantia are located in northern Spain, near Soria (234 km/145 miles northeast of Madrid) in the wooded hills of the Sierra de la Demanda. A rich Museum Numantinus can be found in Soria (Fodor's 99, *Spain*).

15. Carlos Fuentes. *The Buried Mirror. Reflections on Spain and the New World* (Boston 1992).

16. Scipio the Younger (or Minor) was adopted by the eldest son of Scipio Africanus Major. Scipio the Younger served in the army in Spain (151 BC) and as consul (147 BC). Fighting in Africa—like his grandfather—he terminated the Third Punic War capturing and razing Carthage (147 BC). He was censor in 142 BC and consul again in 134 BC when he was sent to Spain to terminate the Numantine Wars in 133 BC. See, *The Columbia Encyclopedia* (Fifth Edition, Columbia University Press 1993).

17. Leonard A. Curchin. *Roman Spain: Conquest and assimilation* (London, 1991). Mike Dobson. *The Roman camps at Numantia, Spain. A reappraisal in the light of a critical analysis of Polybius' discourse on the Roman army.* Ph.D. Thesis, University of Exeter, 1996. Internet abstract at http://www.ex.ac.uk/pallas/mike/abstract.html.

18. *Appian's Roman history*, 4 volumes, Loeb Classical Library, Translated by Horace White, Harvard University Press (Cambridge, MA 1982).

19. University of Texas. LANIC: Latin American Network Information Center. Castro Speech Data Base. www.lanic.utexas.edu/project/castro/db/1990.html.

20. James A. Michener. *Iberia: Spanish travels and reflections* (New York, 1968) offers a definition of *pundonor*: "...it has been left to Spain to cultivate not only the world's most austere definition of honor, but also to invent a special word to cover that definition. Of course, Spanish has the word honor, which means roughly what it does in French, but also the word *pundonor*, which is a contraction of *punto de honor* [point of honor]." See also: John A. Crow. *Spain: The root and the flower. An Interpretation of Spain and the Spanish people* (Berkeley, 1985); Adrian Shubert. *The land and people of Spain* (New York, 1992).

CHAPTER 13

An Unknown Virus

Infectious diseases are the constant companions of our lives.
Charles Nicolle (Nobel Laureate)

The conquest of the Caribbean islands was a deadly enterprise. Explorers were enchanted first by the warm scented sea breezes, the powder-white beaches bordered by palm trees along the turquoise-blue Caribbean Sea, the mountains, the flatlands, and the jungles of the interior. But these alluring geologic features hid lethal microbes fatal to the explorers. For centuries, yellow fever, dysentery, and typhoid fever ravaged the lives of sailors, settlers, natives, and slaves. Colonial churchyards and cemeteries became the final resting places for thousands of men, women, and children, colonists and sailors alike, who died from malaria and yellow fever— "yellow Jack," the mysterious black vomit—that would appear and then vanish again without reason.

On the European continent, the need to protect the health, welfare, and productivity of their colonies forced the European imperial powers to create tropical medicine institutes in London, Liverpool, Berlin, Paris, and Antwerp. In contrast, Castro's Revolution saw the need to locally study the diseases of the tropics and created in Havana a tropical medicine institute named after one of Cuba's most distinguished researchers Doctor Pedro Kourí.

"The life of a man lasts two monsoons," went the English saying in India, and this was also true in the British West Indies.[1] The human cost of taming these lands was appalling. Historians of the day recorded the devastation of the colonial populations. According to Richard Dunn, in *Sugar and Slaves*[2]: "Some 12,000 Englishmen came to Jamaica in the first 6 years, yet the population of the colony in 1661 was only 3470." Among British Army garrisons, more than 100 men of each 1000 soldiers died every year of tropical diseases. The situation was notoriously worse for the slaves. By 1790, Barbados, Jamaica, and the Leeward Islands had imported some 1,230,000 slaves from Africa, and yet the total black population was only 387,000 that same year.[3]

In the Hispano-American colonies, and in Cuba in particular, the situation was no different. During the years 1648 and 1649, a terrible epidemic of yellow fever—a viral disease transmitted by mosquitoes—devastated these Spanish colonies causing the largest number of deaths in the cities of Havana and Veracruz. Yellow fever was the main obstacle to the creation of the Panama Canal.[4] Finally, Carlos Finlay, a Cuban physician demonstrated that yellow fever is transmitted by mosquitoes and can be controlled.[5]

Spain would not surrender Cuba and Puerto Rico, the last two colonies of its once formidable colonial empire in the New World, and devoted major efforts to keeping Cuba under the empire's bicephalous eagle and Spain's gold and crimson flag until 1898, when the United States won Cuba's independence at the conclusion of the brief Spanish-American War.

Late in the nineteenth century, between 1881 and 1894, Spain encouraged large-scale civilian migration to the islands, and about 250,000 migrants from the Spanish mainland and the Canaries arrived in Cuba. Also, from 1791 until 1867, Spain encouraged the development of the sugar industry, and Cuba brought in nearly 800,000 Africans as slaves. By 1870, about 40% of the world's total sugar production came from Cuba's sugar cane fields.[6]

Perhaps because of Cuba's riches, Spain reacted with unprecedented harshness to the movements in favor of autonomy on the island. During the Cuban Wars of Independence (1868–1898), Spain sent to Cuba over 400,000 troops, in contrast to some 47,000 soldiers sent to all the Hispano-American colonies during the numerous wars of independence in Latin America (1810–1820).[7]

During wartime, at least one-quarter of the soldiers died from yellow fever and malaria, and virtually every one of them fell ill with fevers and dysentery. Tropical diseases became an ally of nationalist Cubans during Cuba's long fight for independence from Spain.

The Spaniard Santiago Ramón y Cajal, who received the 1906 Nobel Prize in Medicine, was one of history's most distinguished neuroscientists. Drafted as a young physician, Cajal was sent to Cuba, where he experienced first-hand the terrors of the tropics. At 21 years of age, Cajal was incorporated as physician second rank into the Military Health Regiment in Burgos in 1873. A year later he wrote, "Having received my documents and embarkation salary, I went to Cadiz, where I would board the steamer *España* bound for Puerto Rico and Cuba."[8]

Cajal's first assignment was at Vista Hermosa's infirmary deep in the jungle, where he soon caught malaria. He was sent to the Military Hospital in Puerto Principe, the capital of the Cuban province of Camagüey, for a long convalescence. After his recovery, Cajal was sent to the San Isidro's infirmary in the *trocha del Este*—the eastern fortified trail that traversed Cuba from Morón to Júcaro to prevent rebel troops movements. One year later, in June 1875, he was repatriated to Spain. Cajal returned to his home country "seriously sick, prostrated and melancholic, victim of malarial cachexia." A few months later he coughed up blood and was diagnosed with an episode of hemoptysis from tuberculosis. But it was equally possible he was also suffering from the "black vomit" of yellow fever. Now, 115 years after Cajal almost succumbed to the mysterious diseases of the tropics in Cuba, the members of the Mission appeared to be confronting a tropical disease of unknown origin, supposedly viral.

PEDRO KOURÍ TROPICAL MEDICINE INSTITUTE, HAVANA, CUBA

The following morning Tuesday, May 18, 1993, the Mission and ORBIS teams were taken by bus to the Pedro Kourí Institute, where we were presented with detailed information about the possibility that the causative factor for the epidemic of optic neuropathy was a virus.[9] The presentation was to be led by Dr Pedro Mas Lago.

The respected virologist's small frame and slow, purposeful movements gave an impression of frailty. This morning at the Institute Dr Mas spoke with the quiet assurance of many years of laboratory experience in virology. He adjusted his thick glasses and began in a raspy voice:

"On April 8, 1993, we isolated a viral agent from samples of cerebrospinal fluid (CSF) from patients with epidemic neuropathy. CSF samples had been inoculated in tissue cultures of green monkey kidney cells (Vero cells). In a small percentage, limited alterations of the cells consistent with cytopathogenic effect were observed. Intracerebral and subcutaneous inoculation in newborn Balb/C mice produced hyaline muscle changes, lesions of the enteric mucosa, and encephalitis, suggesting that the agent is a probable enterovirus."

During the preceding 2 months of the epidemic, Dr Mas and his colleagues at the Kourí Institute had conducted viral studies in three laboratories specialized in enteroviruses, herpes viruses, and retroviruses. Only the first laboratory returned positive results.

The name "enterovirus" (from the Greek *enteron*, meaning intestine) indicates that these ribonucleic acid (RNA)-containing viruses are transmitted via human waste. Contagion occurs by the fecal–oral route as a result of poor hygiene and transmitted through the hands of people who fail to wash thoroughly after defecation. Enteroviruses usually cause foodborne illnesses, but they may attack the nervous system also. One of the best-known enteroviruses is poliovirus, the cause of polio, or paralytic poliomyelitis.[10]

When the nervous system suffers an enteroviral attack, there is an immediate and violent inflammatory response with fever and white blood cells mobilized in large numbers toward the CSF. In this battle, meningitis is produced to eliminate the viral invader. The inflammatory response affects the brain and the spinal cord after the invasive virus targets the motor neurons and causes paralysis.

Dr Mas' presentation concluded and was received silently by the members of the Mission. It was evident that the work of virus isolation and identification was quite advanced, but no one in the group had the expertise in virology required to judge the findings.

Could the epidemic be caused by a virus—as originally suspected by Admiral Young?

At first glance it seemed impossible. The epidemic was not spreading as a viral transmission in traditional ways, between spouses, from parents to children, and among coworkers at job sites or classmates at schools.

None of this was present in Cuba, where cases seemed to occur here and there, each one unrelated to the others. Even the best epidemiologic detectives had failed to establish that any contact had occurred between patients. Patients had never been together at a government-organized political rally or had traveled in the same crowded bus. After all, with the *período especial*, nothing moved in Cuba.

Also compelling was the absence of signs of infection. The disease advanced without fever, with normal white cell counts in blood, and without alarming increases in the number of white cells in CSF.

There was also the unexplained resistance to the disease in pregnant women or in children—who, because of the closeness of their classrooms and the high degree of personal contact and touching associated with their games, are a favorite target of all viruses.

Finally, as far as we could remember, none of the recognized enteroviruses was known to affect the optic nerve with the selectivity we had observed in patients affected by the Cuban epidemic.

But even with all these facts, we could not take this viral isolation lightly, as the number of cases continued to rise with increasing vigor.

Fortunately, the virologists were scheduled to arrive in Havana the next day.

ENDNOTES

1. Quoted by John D. Spillane. *Tropical Neurology* (London, 1973).
2. Richard S. Dunn. *Sugar and slaves: The rise of the planter class in the English West Indies, 1624–1713* (Chapel Hill, NC, 1972).
3. Richard B. Sheridan. *Doctors and Slaves. A medical and demographic history of slavery in the British West Indies, 1689–1834* (Cambridge, 1985); Hugh Thomas. *Cuba or The Pursuit of Freedom* (New York, 1998).
4. David McCullough. *The path between the seas. The creation of the Panama Canal, 1870–1914* (New York, 1977); Frederic J. Haskin. *The Panama Canal* (Garden City, New York, 1913).
5. José López Sánchez, *Finlay, El hombre y la verdad científica* (La Habana, 1987).
6. SW Mintz, *Sweetness and power. The place of sugar in modern history* (New York, 1985).
7. C. Schmidt-Nowara, Imperio y crisis colonial, en: J Pan-Montejo (coord.), *Más se perdió en Cuba. España, 1898 y la crisis de fin de siglo* (Madrid, 1998); Hugh Thomas (op. cit.).
8. A. Albarracín. *Santiago Ramón y Cajal o la Pasión de España* (Barcelona, 1978).
9. The Kourí laboratory also developed reagents for the micromethods used by the Cuban SUMA diagnostic equipment for diagnosis of human retroviruses. More than 12 million tests had been performed in Cuba, including blood banks and pregnant patients. Moreover, this laboratory explored the worrisome possibility of finding unknown human retroviruses imported from Africa. By 1992, only 3 cases of the African strain of HIV type 2 had been discovered in soldiers returning from Guinea-Bissau, and only 9 cases of HTLV-I/II had been found in almost 20,000 samples.
10. Enteroviruses include poliovirus, the agent of paralytic poliomyelitis; Coxsackie virus; echovirus; and hepatitis A virus.

CHAPTER 14

The Virologists' Arrival: Havana–Cuba
10:00 am, Wednesday, May 19, 1993

Periodically, arising from the swarm of microbes that traveled with people, since time immemorial, came epidemics that nearly wiped out populations, almost overnight.
Gina Kolata[1]

The arrival of the virologists—including a Nobel laureate—received full attention from the media. The news of the virus isolated by the Cubans had leaked to the press in the United States causing growing concerns that the disease could spread to Miami. The news also fueled the Cuban suspicions that the virus was of hostile American origin. From Havana, Pascal Fletcher reported as follows for Reuters[2]:

US NOBEL WINNER HELPS CUBANS WITH EPIDEMIC

Havana—Dr Carleton Gajdusek, a US Nobel Prize winner in medicine, flew into Havana to join an international team of specialists helping Cuban doctors track down the cause of an epidemic nervous disease. Foreign diplomats in Havana said the arrival of American doctors, some from the National Institutes of Health (NIH), could reflect concern among US authorities that the disease could spread to the United States, just 90 miles away.

But Foreign Minister Roberto Robaina, echoing public suggestions already made by other senior Cuban officials, said Cuba could also not rule out the possibility that the disease had been deliberately introduced. "Enemy action cannot be ruled out," he told the Cuban news agency **Prensa Latina**, *though he said there was no evidence to support this so far. Roughly 65,000 Cubans visited or relocated to the United States in fiscal 1991, the most recent figures available from the US Immigration and Naturalization Service. But no cases of the mysterious ailment, which often affects eyesight, have been reported at the leading eye institute in Miami or at local eye doctors, said Cynthia Diaz, a spokes-woman for the Bascon Palmer Eye Institute, affiliated with the University of Miami.*

The team of top virologists included Dr Carleton Gajdusek and two of his closest collaborators, Drs Paul W. Brown and David M. Asher. I was asked by Minister Teja and Dr Márquez to be part of the welcoming party.

Cuban Blindness. DOI: http://dx.doi.org/10.1016/B978-0-12-804083-6.00014-X

We returned to the José Martí International Airport to greet the new members of the PAHO Mission to Cuba team. Again, full media coverage awaited the visitors, Gajdusek was interviewed, and then the group joined the rest of the Mission in the Biocaribe Hotel.

★ ★ ★

Dr David M. Asher was a Harvard medical graduate with a brilliant scientific career and publications.[3] His graying beard and soft-spoken manner gave him the demeanor of a scientist from the early twentieth century. He was a dedicated worker and a serious, meticulous researcher. He was a pediatrician by training and had developed his career as a virologist at Dr Gajdusek's laboratory at the NIH. He was fluent in Russian, and, as a member of the US–USSR Health Exchange Team, went to the Ivanovsky Institute of Virology in Leningrad, where he studied an unusual form of epilepsy prevalent in Siberia, first described by Kozhevnikov as *Epilepsia partialis continua*.[4] Most recently, he had participated in the study of an epidemic outbreak of encephalitis in cows in Britain—the so-called mad cow disease—which riddled the cow's brain with sponge-like holes (spongiform encephalopathy) and was caused by scrapie virus from sheep.[5] This epidemic nearly destroyed Great Britain's cattle industry. More than 4.7 million cows were put to death and incinerated at high temperatures.

The cows had become infected when bone meal obtained from scrapie-infected sheep carcasses had been added to cattle feed as a source of calcium and proteins. In this way, humans managed to turn vegetarian cows into passive carnivores, and the virus made the unnatural jump from the ovine species to the bovine species.

The European scare stemmed from the reported increase in Britain of human cases of a new variant of Creutzfeldt–Jakob disease (CJD), one of the transmissible spongiform encephalopathies of humans. This would indicate that the rugged scrapie virus had once again jumped species, this time from the bovines to the humans—assumed to have been infected by eating beef from cattle harboring the agent of the prion disease; this was enough to turn many meat-eaters into vegetarians.

Dr Asher would contribute to the Mission team his extensive laboratory experience in virology and his international expertise. These would be useful in evaluating the Cuban viral isolate. Furthermore, he would be using Russian to communicate with the Cuban colleagues, since he spoke no Spanish and few Cubans spoke fluent English.

Dr Paul W. Brown, another Harvard graduate, had also been a long-time collaborator of Dr Gajdusek and an excellent traveling companion. While waiting in Miami for the flight to Cuba, he and Dr Gajdusek recalled an extraordinary journey many years ago from Teheran, Iran, to Kabul, Afghanistan, by way of Mazar-e-Sharif, after an overnight stop in Heart—names that will become familiar to the public following the Iran and Afghanistan wars yet to come.[6]

Dr Brown was the medical director of Dr Gajdusek's laboratory at the NIH. A graduate from Johns Hopkins School of Medicine, he completed his medical residency at Hopkins' Osler Medical Service and at the University of California Medical Center. Dr Brown was fluent in French and had worked in Paris as *Chargé de Recherche* at INSERM—the French equivalent of the NIH—and as Medical Consultant to the American Embassy in Paris. He was a visiting scientist at the Hôpital de la Salpêtrière, the famous neurologic hospital, where I made my first strides at becoming a neurologist. Dr Brown had been awarded the Prix Léopold Talbot of the French Veterinary Academy and the Commendation and the Outstanding Services Medals of the US Public Health Service.

Dr Brown was an expert in the slow virus infections of the nervous system, in particular CJD and "mad cow disease," serving as consultant to the World Health Organization/Pan American Health Organization (WHO/PAHO) and the European Community on CJD surveillance.[7] The visit to Cuba was taking him away from his current work on "mad cow disease" in Europe.[8]

★ ★ ★

In 1957, Gajdusek was 33 years old when he visited Papua and New Guinea looking for kuru, a word that means "to be afraid" or "to shiver" in the language of the Fore people of New Guinea. Old photographs show Dr Gajdusek as a scrawny youngster with short hair and a timid smile, wearing a white T-shirt and canvas shorts. He was now an older man of imposing, almost pontific, presence but the smile remained gentle and timid, and his gestures revealed his simple gifts for hospitality and generosity learned in a lifetime spent among primitive peoples.

During his first visit to the core of the kuru epidemic, he was impressed by this stone-age tribe, whose women were "cooking and feeding their children the body of a kuru case." Eventually Dr Gajdusek would demonstrate that the slow virus causing kuru was transmitted by cannibalism. This earned him the 1976 Nobel Prize in Physiology and Medicine for the discovery of new agents of human disease.[9] He called these agents "unconventional or slow viruses," known today as prions.[10]

A productive insomniac, Dr Gajdusek was a voracious reader and a prolific writer with more than a thousand published scientific articles to his name. His diaries, handwritten in small careful penmanship, contain meticulous, incisive, and irreverent observations on people and geography. His trip to Cuba would also become part of these memoires.[6]

Daniel Carleton Gajdusek was born on September 9, 1923, in Yonkers, a New York suburb, to a Slovak farmer father, who prospered as a master butcher, and a Hungarian mother. His only brother is poet and writer Robert E. Gajdusek. From his father he inherited an indefatigable energy, a gift for conversation, and a *joie de vivre*. From his mother and her family he inherited his love of books, science, and study. His love of languages resulted from the influences of the multilingual Rumanian and Hungarian migrant community in which he grew up. Dr Gajdusek was a gifted polyglot, fluent in English, German, French, Spanish, Russian, Slovak, Dutch, Neo-Melanesian, Persian, and Bahasa-Indonesian, and he was able to make himself understood in several languages of the numerous linguistic groups of Papua and New Guinea, Melanesia, and Micronesia.

Dr Gajdusek attended the University of Rochester and graduated *Summa cum Laude* in Biophysics, received his MD degree from Harvard University, and did his residency in pediatrics at the Babies Hospital of Columbia Presbyterian in New York and at the Cincinnati Children's Hospital. In 1948 he moved to California as a postdoctoral fellow at Cal-Tech, where he worked under future Nobelists Linus Pauling and Max Delbrück.

From the beginning of his career, Dr Gajdusek's research interests were focused on the infectious diseases of children. His mentor was Dr Joseph E. Smadel at the Walter Reed Army Institute for Research in Washington DC and future associate director of the NIH. "Dear Joe" saw the enormous potential of this young man beyond his eccentric surface, and with unfailing faith, he fought bureaucracy and red tape to make Gajdusek's discoveries possible. He supported him while he conducted research on *Pneumocystis carinii* infection, a viral disease occurring in children; he went looking for viral diseases among Okinawan migrants colonizing the jungles along the Rio Guapay in Bolivia and in the Peruvian Amazon basin; the Tarahumara Indians in Sonora; and, the Chihuahua in Mexico. Dr Gajdusek described acute infectious hemorrhagic fevers and

mycotoxicoses in Asia and in the former Soviet Union, performed serological surveys of poliomyelitis, herpesvirus infection, mumps, arthro-
pod-borne viral diseases, leptospirosis, syphilis, and rickettsial diseases, in
particular Q fever, in the Middle East, mainly in Afghanistan, Iran, and
Turkey.[11]

During these years, Dr Gajdusek developed a productive collaboration
with Dr Marcel Baltazard of the Institute Pasteur in Teheran, Iran, where
they studied Pasteur's disease—rabies—transmitted by ferocious bites of
rabid desert wolves, and with Dr Baltazard tested the use of the antirabic
serum.[12]

The year 1955 saw Dr Gajdusek working in Australia under Sir Frank
Macfarlane Burnet, Director of the Walter and Eliza Hall Institute of
Medical Research in Melbourne. Dr Burnet would be awarded the 1960
Nobel Prize for his work in autoimmunity and self-tolerance.[13]

Kuru became the only theme in Dr Gajdusek's life, an obsession
dominating days and nights of frantic activity during the 10 months of
uninterrupted field work in the mountain ranges of New Guinea with
Estonian-born Dr Vincent Zigas as his lone colleague fighting the opposi-
tion of the Australian government.[14]

Kuru, the primitive natives, the rugged steep mountains, and the isolation of the jungle—the mystery of it all—formed the perfect combination.
Kuru was the challenge he had prepared for all his life. And he would not
stop until he had solved the riddle and had received the Nobel Prize in
1976 for his discovery.[9]

★ ★ ★

That evening, the visiting virologists met their Cuban hosts at an
event held in the Biotechnology Institute auditorium. The leading per-
sonalities of the *Grupo Operativo* were already in the front line, alongside
the members of the Mission. Fidel Castro then entered the auditorium
to the applause of the standing audience. Dr Miguel Márquez introduced
the Nobel laureate:

"Comandante," he said breathlessly, "this is Dr Daniel Carleton Gajdusek
from the NIH in Washington, winner of the Nobel Prize in Medicine in
1976."

Fidel embraced Dr Gajdusek warmly, and the two men exchanged
greetings in Spanish.

It was a historic moment. After more than 30 years of isolation and
embargo, the United States had responded to the plea of the Cubans by

sending to Cuba the most distinguished scientist in the United States in the field of virology and the nervous system.

Castro then greeted Dr Asher and Dr Brown, and following his custom, shook hands with all the members of the Mission, greeting them in the Cuban fashion by their last names "Llanos… Caballero… Silva… Muci… Román…"

Then the Comandante took his seat, and the meeting began.

The Cubans had been able to isolate a virus in 47 of 49 samples of cerebrospinal fluid (CSF) from patients with neuropathy. In every instance the CSF samples had been normal, failing to show increases in the number of cells counted in each cubic millimeter and having the usual amounts of protein and sugar. In some cases repeated isolations of the virus had been achieved despite clinical improvement with the intravenous vitamin treatment.

One by one, the laboratory researchers presented each of the small pieces of the puzzle and the techniques used for tissue cultivation of the virus and inoculation into newborn mice and in rabbits. Striking electron microscope photographs revealed the unmistakable images of small viral particles.

Then the molecular biologists presented the results of the tests they had used to detect the deoxyribonucleic acid of the virus in CSF and blood. They had used a new test called *polymerase chain reaction* to look for the enterovirus' signature. The Cubans had produced a genetic library with a sequence obtained from a fragment of the virus message for the protein VP1. The virus isolated from patients had fingerprints that matched, in 91% of the features, an enterovirus, more precisely Coxsackie virus subtype A9.

The Cubans had gone one step further. By means of molecular techniques they had cloned some of the proteins of the virus to develop a blood test to be able to measure the human response against a probable infection with this agent to determine the approximate time of appearance of this virus in Cuba.

Lights were turned on in the auditorium, and Minister Teja asked Dr Gajdusek to sum up the massive amount of information provided in the long briefing. Dr Gajdusek obliged. He walked to the podium, took his glasses off, looked briefly at Fidel Castro, and then addressed the audience in a barely audible voice. He thanked the Cuban leader for his hospitality, congratulated the Cuban scientists on their hard work, and spoke admiringly of their diligence and devotion to scientific method.

Then—his voice slowly rising—he urged his Cuban colleagues to never forget the human source of their scientific samples. They must think always of the patient, he told them. They must never mistake the human being as a scientific cipher. "Do not practice virology in a void!" he urged them.

Looking again at the notes in his hands, Dr Gajdusek reminded his colleagues of an important historical precedent.

"Never—and I mean *never*—in the history of poliomyelitis, the most typical of the neurologic infections by enteroviruses, were we able to isolate the virus from the spinal fluid in more than 10% of the cases," he told his audience. "But you are having isolations of an enterovirus, a virus of the same family as poliovirus, in 96% of the cases."

He shook his papers at the audience. "Ninety-six percent! And in patients who have not one single inflammatory cell in the spinal fluid."

Dr Gajdusek waited, to make sure his audience grasped the full meaning of his outburst. Then he said, in a calm voice, "Search no more. You have isolated a virus in 96% of your cases. This is the cause of the disease. And *if* this virus really exists, we will need to rewrite a whole chapter on the virology of the nervous system!"

The audience remained totally silent as Dr Gajdusek left the podium and returned to his seat. Then, slowly at first, the applause began. Soon the entire audience was clapping. Even Fidel Castro, the master orator himself, stood up to congratulate the scientist on his brave performance.

ENDNOTES

1. Gina Kolata. *Flu. The story of the Great Influenza Pandemic of 1918 and the search for the virus that caused it* (New York, 1999).
2. Pascal Fletcher, U.S. Nobel winner, helps Cubans with epidemic. *Reuters News Agency. Washington Post* (20 May 1993, p. A7). The report continued as follows:
   US Nobel Winner Helps Cubans with Epidemic (cont.)
   Cuban experts have admitted they are puzzled about the precise origin of the disease, which has affected more than 25,000 people in the communist-ruled Caribbean island and is still spreading. Cuban experts have initially attributed the outbreak to a possible combination of vitamin deficiency and an unidentified toxin or virus.
   Cuba, on the grips of a severe economic crisis, has appealed for international scientific help to pinpoint the exact cause of the illness, which Cuban doctors describe as an "epidemic neuropathy," a disease of the nervous system that affects eyesight and other parts of the body.
   "We have to learn what you have learned already and what you are doing," Dr Gajdusek, the 1976 Nobel Prize winner in Medicine, told Cuban Health Ministry officials at Havana airport. "There are many different types of neuropathy," Dr Gajdusek said. "One of the commonest…[affects people who] drink methyl alcohol…then there are other causes that are autoimmune, there are genetic causes," he added. He went on: "Your own work will give us an idea of the possibilities." Dr Gajdusek said the two

colleagues with him, David Asher and Paul Brown, had worked on neuropathies found in Jamaica and other parts of the Caribbean, and in Africa, China and Siberia. He stressed that while neuropathies were widespread, pinning down the precise cause in each case was much more complex.

Dr Gajdusek is chief of the Laboratory of Central Nervous System Studies, part of the National Institute of Neurological Diseases at NIH. The three will join other US experts of the private US eye care organization ORBIS and five other mostly Latin American doctors, who have all been brought to Cuba by the Pan American Health Organization to help in the fight against the epidemic. Cuban President Fidel Castro has taken a personal interest in the medical inquiry, meeting Cuban and foreign specialists every day to receive an update on its progress.

Cuban hospitals are providing 20,000 extra beds to deal with the victims of the epidemic of the disease, which has affected all of the Island's 14 Provinces. Patients complained of disruptions of their eyesight, problems of muscular movement, lack of coordination and painful cramps, or all of these symptoms together.

3. David M Asher, et al. Experimental kuru in the chimpanzee. In: W. Montagna, WP McNulty (eds.), *Nonhuman primates and human disease* (Basel, 1973); E Beck, PM Daniel, DM Asher, DC Gajdusek, CJ Gibbs Jr. Experimental kuru in the chimpanzee: A neuropathological study. *Brain* 1973; 96(Part 1):441–462; CJ Gibbs Jr, DC Gajdusek, DM Asher, et al. Creutzfeldt-Jakob disease (spongiform encephalopathy): Transmission to the chimpanzee. *Science* 1968; 161:388–389.

4. David M Asher. Translation from the Russian of articles by AY Kozhevnikov: A peculiar form of cortical epilepsy (Epilepsia corticalis sive partialis continua) and LI Omorokov: Kozhevnikov's epilepsy in Siberia. In, F. Anderman (ed.), *Chronic Encephalitis* (Boston, 1991). FW van der Waals, DM Asher, et al. Post-encephalitic epilepsy and arbovirus infections in an isolated rain forest of Central Liberia. *Tropical and Geographical Medicine* 1986;38:203–208; DM Asher. Movement disorders in Rhesus monkeys after infection with tick-borne encephalitis virus. *Advances in Neurology* 1975; 10:277–283; DM Asher et al. Antibodies to HTLV-I in populations of the Southwestern Pacific. *Journal of Medical Virology* 1988; 26:339–351; A Taller, DM Asher, et al. Belgrade virus, a cause of hemorrhagic fever with renal involvement in the Balkans. *Journal of Infectious Diseases* 1993;168:750–753; BK Johnson, DM Asher, et al. Long-term observations of HIV-infected chimpanzees. *AIDS Research and Human Retrovirus* 1993; 9:375–378.

5. David M. Asher. Recommendations concerning the risk of bovine spongiform encephalopathy in the United States. *Journal of the American Veterinary Medical Association* 1960; 196:1687; CJ Gibbs Jr, L Bolis, DM Asher, et al. Recommendations of the international round table workshop on bovine spongiform encephalopathy. *Journal of the American Veterinary Medical Association* 1992; 200: 164–167.

6. D. Carleton Gajdusek. Consultation on Epidemic of Acute Optic Neuropathy, Havana, May 16 to May 22, 1993 (personal diary, typewritten manuscript, by courtesy of the author).

7. Paul W. Brown et al. Creutzfeldt-Jakob disease in France. *Annals of Neurology* 1979; 6:430437; PW Brown. Epidemiology of Creutzfeldt-Jakob disease. *Epidemiological Records* 1980; 9180; 2:113–135; PW Brown, et al. "Friendly fire" in medicine. Hormones, homografts and Creutzfeldt-Jakob disease. *Lancet* 1992;340:24–27; PW Brown, et al. Intracerebral distribution of infectious amyloid protein in spongiform encephalopathy. *Annals of Neurology* 1995;38:245–253.

8. Richard Rhodes. *Deadly feasts. Tracking the secrets of a terrifying new plague* (New York, 1997); Sarah Lyall: The British mad cow scare. *NY Times News Service* (4 March 1996); Alan Cowell: European Union and the British mad cow disease. *NY Times News Service* (4 March 1996).

9. D. Carleton Gajdusek, Autobiographical sketch. In: *Les Prix Nobel en 1976* (Nobel Foundation, Stockholm, 1977); R. Bingham, Carleton Gajdusek. Outrageous ardor. In: Ellen L. Hammond (ed.), *A passion to know: 20 profiles in science* (New York, 1984); J. Goodfield. The Kuru Mystery. In, *Quest for the Killers* (Boston, 1985).

10. Stanley B. Prusiner. Novel proteinaceous infectious particles cause scrapie. *Science* 1982; 216:136–144; SB Prusiner, KK Hsaio. Human prion diseases. *Annals of Neurology* 1994;35:385–395.

11. D. Carleton Gajdusek. *Pneumocystis carinii*—Etiologic agent of interstitial plasma cell pneumonia of premature and young infants. *Pediatrics* 1957; 19 (4); Part 1: 543–565 (1957); DC Gajdusek, et al. Serological survey of viral and rickettsial diseases among jungle inhabitants of the upper Amazon Basin. Presence during infancy, childhood and adolescence of antibodies to rickettsiase, to viruses of poliomyelitis, mumps, herpes, PLV group, yellow fever and to the group B Arthropode-borne viruses. *Pediatrics* 1959;23:Part I:121–131; Schaeffer M, Gajdusek DC, et al. Epidemic jungle fevers among Okinawan colonists in the Bolivian rain forest. *American Journal of Tropical Medicine & Hygiene* 1959; 8 (3):372–396; 8(4):479–487; DC Gajdusek and NG Rogers, Specific serum antibodies to infectious disease agents in Tarahumara Indian adolescents of northwestern Mexico. *Pediatrics* 1955;16(6):819–835; DC Gajdusek. Hemorrhagic fevers in Asia: A problem in medical ecology. *Geographical Review* 1955;46(1):20–42; DC Gajdusek. Acute infectious hemorrhagic fevers and mycotoxicoses in the Union of Soviet Socialist Republics. *Walter Reed Army Medical Center* (Washington, 1953); DC Gajdusek. Das epidemische hämorrhagische Fieber. *Klinische Wochenschrift* 1956;34(29–30):769–777; DC Gajdusek, M. Bahmanyar, Sur la Q fever en Iran. *Bulletin de la Societé de Pathologie Exotique* 1955;48 (1):31–32.

12. M. Baltazar, M. Bahmanyar, M. Ghodssi, A. Sabeti, C. Gajdusek, and E. Rouzebehi, Essai pratique de sérum antirabique chez les mordus par loups enragés. *Bulletin of the WHO* 1955; 13 (5):747–772.

13. DC Gajdusek. An "auto-immune" reaction against human tissue antigens in certain acute and chronic diseases. *Nature* 1957;179: 666–668.

14. D. Carleton Gajdusek. Correspondence on the discovery and original investigations on kuru. Smadel-Gajdusek Correspondence, 1955–1958; (Washington, 1976); Kuru Epidemiological Patrols from the New Guinea Highlands to Papua, 1957 (Washington, 1976); *Kuru: Early letters and field-notes from the collection of D. Carleton Gajdusek*. Edited by Judith Farquhar/D. Carleton Gajdusek (New York, 1981). DC Gajdusek. In memoriam: Vincent Zigas 1920–1983. *Neurology* 1982;33:1199–1200; DC Gajdusek, V. Zigas. Degenerative disease of the central nervous system in New Guinea. The endemic occurrence of "kuru" in the native population. *New England Journal of Medicine* 1957;257 (30):974–978; DC Gajdusek and V. Zigas: Untersuchungen über die Pathogenese von Kuru: eine klinische, pathologische und epidemiologische Untersuchung einer chronischen, progressiven, degenerativen und unter den Eingeborenen der Eastern Highlands von Neu Guinea epidemische Ausmaße erreichenden Erkrankung des Zentralnervensystems. *Klinische Wochenschrift* 1958;36 (10): 445–459 (1958); DC Gajdusek, V. Zigas, J. Baker: Studies on kuru. III. Patterns of kuru incidence: demographic and geographic epidemiological analysis. *American Journal of Tropical Medicine & Hygiene* 1961; 10 (4):599–627.

CHAPTER 15

Mariel
Thursday, May 20, 1993

> *At least, 10 per cent of the total population of Cuba was trying to escape.*
> **President Carter**

The entire personnel of the Mission to Cuba left Havana early next morning for the two-and-a-half-hour trip to Pinar del Río. The motorcade of blue minibuses and black Lada sedans was alone on the deserted highway. A few kilometers up, at the outskirts of Havana, on the right-hand side of the road, an arrow pointed north to the port of Mariel.

The name Mariel will be always associated with the diaspora of Cuban people and will cause echoes of despair for some Americans. During April and May 1980, in the last year of his administration, President Jimmy Carter had to confront a massive influx of Cuban refugees. In his Mémoirs, the former US President wrote[1]

> *In the spring and summer I had to deal with a stream of illegal Cuban refugees who began coming to our country. We welcomed the first ones to freedom, but when the stream became a torrent, I explored every legal means to control the badly deteriorating situation. Even so, it was impossible to stop all of them…*
>
> *Because of problems within Castro's regime, at least 10 per cent of the total population of Cuba was trying to escape, and looking for any possible means to come to the United States. At the same time, in Miami and throughout the country, there were tens of thousands of Cuban-American families eager to bring their relatives out of Cuba to join them here. In many cases these people were influential, with the financial resources to pay for their relatives' travel either openly or by a clandestine route.*
>
> *Under pressure at home, Castro began actively encouraging many Cubans to emigrate. Although most of the emigrants were good citizens, we were soon to discover that some were mental patients and criminals…*

From this small port, a small armada of boats ferrying the so-called *Marielitos* to Miami exiled 125,000 Cubans eager to leave the island.[2] Among them were prostitutes, psychiatric patients, and criminals from Cuba's toughest prisons. In Castro's own words, he sent to exile "the lumpen, the worms, the undesirables." The arrival of these exiles on American soil was

Cuban Blindness. DOI: http://dx.doi.org/10.1016/B978-0-12-804083-6.00015-1

one of the factors that led to President Carter's election loss to his successor Ronald Reagan.

Years later, President Bill Clinton's political career was also affected by the *Marielitos*. As a young Arkansas governor, he had to confront the insurrection of thousands of refugees who were being held at Fort Chafee, a detention camp in his state, awaiting a decision on their legal situation. The riots contributed to his defeat in the 1980 gubernatorial elections.

As president, Clinton later dealt with the massive 1994 migration of some 32,000 *balseros*, who had fled Cuba in makeshift boats and handmade rafts. In response, Clinton changed the traditional policy of the United States that welcomed Cuban exiles as political refugees and instead declared them illegal immigrants.[3] The US Coast Guard Service began intercepting the *balseros* and placing them under detention at the US Navy Base in Guantánamo.

The only exception to the new US restrictions against Cuban migration was the medical exemption. There ensued among the detainee population at Guantanamo a true epidemic of self-mutilations and serious self-inflicted wounds as desperate Cubans made a last pitiful attempt to escape their homeland.[4]

★★★

We continued traveling on the wide and empty highway. The flat sugar cane fields were left behind, and slowly the horizon began to rise with the first elevations of the *Sierras del Rosario y de los Órganos*, which eventually form, further west, the majestic *Cordillera de Guaniguanico*. On the hills, forests of tall Royal palms provided the typical Cuban characteristic to the tropical landscape.

In the shade of unfinished overpass bridges connecting to nowhere, groups of *guajiros* waited for a ride. Small farms began to appear. Finally, a billboard sign announced: "You are now entering the *Provincia de Pinar del Río, la Cenicienta de Cuba*," Cuba's Cinderella, abandoned and neglected until the Revolution.

Forest patches of native pines recalled the original name of the province, "pine forest by the river." Further down, a large billboard of Ché Guevara in green fatigues proclaimed "*¡Patria o muerte!*"—Fatherland or Death!

Brick and tile houses and low-cost apartment buildings appeared, and the number of people walking by the roadside increased rapidly, as the highway approached the capital of the province. A road sign honored Tanya *la Guerrillera*, one of the few female heroes of the Revolution.[5]

Finally, the convoy stopped at the door of the Abel Santamaría Regional Hospital at the outskirts of Pinar del Río.

For the next several hours, members of the Mission—following a snack of fresh papaya slices, sausage sandwiches, fruit juice, and the perennial *café Cubano*—were treated to an exhaustive report of statistics related to the outbreak of the epidemic in this province.

We were briefed on details such as the relationship between the mineral content of the soil and the frequency of cases in a given area or between patterns of transportation and the appearance of new cases. Meteorologic data showed variations in temperature, humidity, winds, tides, and hours of sunlight, charted from the beginnings of the epidemic to its current peak. Agricultural data listed the metric tonnage and types of foodstuffs produced locally, and those imported and distributed in the Province, as well as the tonnage of fertilizers and pesticides categorized by types, crops affected, areas sprayed, and amounts used.

Epidemiologists had listed the daily number of medical and dental outpatient consultations separated by specialties. Hospitals had organized discharges by diagnoses and the curves of occurrence of most diseases separated by diagnoses from diarrhea to heart failure. There were even figures concerning the sanitary control of domestic animals—implemented since the 1971 epidemic of African swine fever.

It became gradually clear, however, that nowhere in this vast databank of facts and figures was there a single clue to the cause of the epidemic.

We noted from day 1 that female physicians—a growing majority in Cuba, as it had been in the Soviet Union—were in charge of the clinical and epidemiologic aspects of the epidemic in Cuba and Pinar del Río.

Dr Carmen Serrano Verdura, Tenured Professor of Internal Medicine, was responsible for the clinical care of all patients with epidemic neuropathy in the Province. A small, black woman with fine facial features and a soft, authoritative voice, Dr Serrano was the first physician to discover that the peripheral nerves of hands and feet were also damaged in patients with the epidemic disease.

Dr Mariluz Rodríguez Alvarez, Second-Degree Specialist in Epidemiology, had organized and collected all the information for the databases presented to the Mission members. Dr Rodríguez was a tall blonde woman, with short hair, boundless energy, and an easy smile; she favored bright orange and green blouses and chain-smoked black tobacco "Popular" cigarettes. She spoke fast, and often the tempo of her Cuban speech could hardly keep up with the rapid train of her ideas.

The Pinar del Río team was completed by ophthalmologists Dr Blanca Emilia Elliot and Dr Marta María de la Portilla Castro, who together had defined the eye features of the disease, and otorhinolaryngologist Dr Sonia Novales Amado, who had discovered that hearing deficit was a common feature of the disease affecting mainly the perception of high-pitched sounds.

This team had presented a report to the Ministry of Health, as early as May 1992, indicating the strong relationship between the neuropathy cases and tobacco cultivation. They noticed clustering of cases along the roads in the province and absence of familial cases. This suggested that it was highly unlikely that an infectious agent could have caused this neurologic disease.

The Pan American Health Organization group decided to search for a toxic agent that could have been distributed along the rural roads of the province in these times of famine, when people would eat anything edible, including potato and plantain peels.

It was not farfetched to postulate that, for instance, adulterated industrial oil could have been sold as cooking oil by the distribution chains of the black market. Some years ago, a strange epidemic disease in Spain had occurred exactly in this manner when toxic oil had been peddled to families in remote villages. However, the extent of the current outbreak rendered the adulterated oil hypothesis very unlikely.

Date: Thursday, 5/20/1993, Pinar del Río. I could sense the frustration that had taken hold of the entire Mission team in the search for a cause of the epidemic. The disappointment was produced, in no small measure, by the inconclusive results of the extensive research studies performed by our Cuban colleagues. In a way, we were expressing our solidarity with their failure because they appeared to have explored all possibilities of causative factors.

Yesterday's main event was Dr Gajdusek's masterly argument; he used a simple and logical argument to demonstrate that the Cuban viral isolate, in all likelihood, was **only** a laboratory contaminant. We noticed that the epidemiologic pattern was inconsistent with an infectious etiology, but as a clinician, I should have recognized the fact that there was no cell reaction in the CSF of patients despite repeated viral isolations. In agreement with the lack of an infectious pattern of spreading, Dr Gajdusek's conclusion practically eliminated an infectious cause for the epidemic. But then, what *is* the cause? The Cubans have looked everywhere trying to find a culprit to no avail. We arrived 5 days ago, have been briefed on every imaginable aspect of this disease, examined hundreds of patients in hospitals and nursing homes, and, except for Dr Sadun's crucial observation, the rest of us had contributed very little to the solution. We needed to visit the patients at their homes in Pinar del Río. This would be our last opportunity to find a factor that could have been overlooked by the Cubans.

ENDNOTES

1. Jimmy Carter. *Keeping faith. Mémoires of a president.* (New York, 1982).
2. Cuban Information Archives. Cuban Exodus. www.cuban-exile.com.
3. Saul Landau. Clinton's Cuba policy: A low-priority dilemma. *NACLA Report on the Americas* (May 1993). An agreement in December 1984 on immigration procedures allowed the repatriation of 2700 "undesirables."
4. TC Andrews, DL Cull, JI Pelton, SO Massey Jr, JM Bostwick, Self-mutilation and malingering among Cuban migrants detained at Guantanamo Bay. *New England Journal of Medicine* 1997;336: 1251–1253; L. Eisenberg, The sleep of reason produces monsters. Human costs of economic sanctions. *New England Journal of Medicine* 1997;336: 1248–1250 (1997); GC Román, Epidemic neuropathy in Cuba: A public health problem related to the Cuban Democracy Act of the United States. *Neuroepidemiology* 1998;17:111–115.
5. Haydée Támara Bunke, was a German-Argentinian female spy who worked with Ché Guevara in Bolivia. See, JL Anderson. *Ché Guevara. A revolutionary life* (New York, 1997).

CHAPTER 16

Hunger Everywhere
Pinar del Río, May 20, 1993

Theoretically, socialism nationalizes richness. In Cuba, a strange perversion of this practice led to the nationalization of misery.
Guillermo Cabrera Infante

Finally, we met the patients. As in Havana, at the hospital in Pinar del Río we found victim after victim with identical symptoms and eye signs.

Blindness always began following rapid weight loss. Pudgy patients became skinny, and patients who were already thin were reduced to skin and bones. Patients also experienced the urge to urinate frequently, awakening to urinate several times during the night. They experienced high-pitched sounds in the ears, which was accompanied by loss of hearing. Those stricken also had the problem of *los vapores*,[1] a feeling like hot steam scalding the soles of the feet, the fingertips, and the palms of the hands. It was worst at night, keeping the patients awake as they were unable to find a position that would relieve the pains; they would walk trying to forget the pain or take a handful of sleeping pills in an effort to break the suffering.

The numbers of patients were staggering, and there was no doubt that the epidemic was taxing the Cuban medical system badly. Temporary beds had been installed everywhere in the hospital, and supplies were beginning to fail.

The Mission team had been eager to visit the homes of the first optic neuropathy victims. In accordance with our wishes, our Cuban hosts escorted us through the old streets of Pinar del Río.

At the *dispensario* (dispensary) in the Asiento Viejo neighborhood of Pinar del Río, the young doctor in charge appeared fit and trim. The only incongruous detail was his starched and ironed yellow shirt that looked a couple of sizes too large. He had diagnosed more than two-dozen cases of blindness among the 200 families assigned to him. No children, teenagers, or young women with kids had been affected among his patients and neighbors. Most of the families under his care were middle-aged and

Cuban Blindness. DOI: http://dx.doi.org/10.1016/B978-0-12-804083-6.00016-3

older persons with grown-up children no longer living in Pinar. When the epidemic began, he went on a house-to-house search looking for people who had noticed any problems with their eyesight. The patients soon revealed to him the basic ABC of the presenting symptoms of the epidemic: A, *ardor* or *deslumbramiento*, a painful glare caused by the bright tropical sun; B, *visión borrosa*, blurred vision; and C, *ceguera*, or blindness. Once he confirmed that the patients were unable to read the 20/200 letters on his Snellen's pocket card even with their glasses, he sent them immediately to the Abel Santamaría Regional Hospital to be examined and treated by the specialists there. He was seldom wrong in his diagnoses. He proudly told us that at least two-thirds of his patients had regained enough eyesight with treatment to be able to return to work.

"What do you think is the cause of the epidemic?" I asked him, point-blank.

He looked at me with a slightly ironic grin. "Lack of food!" he replied swiftly and continued on a rapid-fire explanation, "This may be an exotic etiology, a very rare diagnosis for you doctors practicing medicine in countries where you spend most of your days telling your patients 'Go on a diet, lose weight, do not eat this, do not eat that.' For us in Cuba this *período especial* has not been a new year's resolution; something like—'In 1992, we will go on a diet'—because when you begin a diet you *know* that food is available and plentiful. I imagine diet is just a question of will, of keeping away from the freezer, to avoid buying the forbidden ice cream, to walk away from the cafeteria. For us it was a gradual almost furtive removal of food from the *bodegas*.[2] This week there is no beef, next week no chickpeas, next the tomatoes are gone, and suddenly you realize that you are always hungry and thinking that for next week they have announced that plantains will be arriving, that the harvest was excellent, and then you notice that for more than 6 months you have been eating only *arroz y frijoles* every single day and you cannot remember when was the last time that you had pork, or fish, or eggs, or milk, or butter, or cheese. When I noticed that I had dropped 10 kilos and that there was no source of vitamins in the official diet, I self-prescribed one vitamin pill every day. That is how I was spared from the epidemic in contrast to many of the *médico de la familia* colleagues, but I am still a good 10 kilos below my normal weight."

He stopped abruptly, slightly embarrassed at his flare-up and then said to us with an apologetic smile, "Well, you are here to see patients, not to listen to the tirades of a country doctor. Let's go."

The buildings were of relatively recent construction and were airy and well shaded by tall almond trees. Balconies opened to a central patio that, instead of lawn and garden, featured a patch of corn, climbing beans, lettuce, cassava, and the typical elephant-ear leaves of *malanga*.[3] In a far corner, a small pen enclosed a piglet, symbolic sign of hope for a future communal *caldoza*.[4]

The homes were small, modest, clean, sparsely furnished, and decorated with party slogans and photographs of the three great heroes of the Revolution—Fidel Castro, Camilo Cienfuegos, and Ché Guevara.

People were hospitable, talkative, quick to smile, willing to collaborate, and grateful for the attention of the physicians who had come from abroad to help them.

We inspected every corner of the small apartments: the dining-living rooms, bedrooms, tiny bathrooms, and kitchens. We found in every household a television, a freezer, electric fans, two recliners, and a rocking chair.

We looked for toxins in the kitchen pantries, refrigerators, and closets, thinking that, perhaps, the contaminant was something so quotidian and obvious that the Cubans had missed its significance.

What we met, though, was poverty and scarcity. The simple lack of food was overwhelming. Everywhere empty freezers sat unplugged or contained nothing but a few potatoes, a jar of cold water, and some medicines. Metal containers designed to hold sugar, rice, and coffee were often empty, too.

Meat, pork, chicken, dairy products—all of which had once been supplied by the government—had almost completely disappeared from the ration cards. Milk was available for children below age 10 years. Eggs were rationed to one egg per person every 2 weeks. Cooking oil was rationed to one small bottle every 2 months, and the quality was so poor that the smelly mixture filled with sediments was often discarded for fear of intoxication.

Almost everyone was barely surviving on the ration cards, with one daily serving of rice and beans (1 pound of beans and 5 pounds of rice per person per month) and one or two small bread rolls daily per person.[5]

In an effort to combat hunger, people were growing anything they could, everywhere, including the lawns and other available patches of dirt. Gardens of corn, cabbage, cassava or *yuca*, sweet potatoes, yam, *malanga*, and beans could be seen where flower beds used to be. Thus for a few Cubans sometimes fresh vegetables were available.

Fresh fruit was scarce, and tomatoes—cultivated for export—almost never appeared in the *bodegas*. The Cuban citrus plantations exported their entire harvest and even tropical fruits, such as guavas and mangoes, were rarely seen, since the small farmers' markets had been closed.

Fresh fish was just a memory. Even though a generous sea surrounds Cuba, the island's entire seafood production was earmarked for export. Sometimes the *picadillo*, a sort of hamburger mix distributed by the government, contained fishmeal instead of meat. Most often it consisted of soybean meal—sold to Cuba at discounted prices by China—mixed with ground meat and meat byproducts. This *picadillo* had such a pervasive flavor of canned dog-food that many found it intolerable.

Late in the *periodo especial*, even Cuban coffee, on which the average citizen subsisted, was becoming a luxury that few could afford. In many households the only products that were abundant and could be had on a regular basis were brown sugar, the cheap "Popular" brand cigarettes, and the weekly allowance of rum.

Cubans, desperate to feed their families, were developing unusual recipes to fight hunger and fill their stomachs. Members of the Mission were told one story, told only half in jest, about a sudden bonanza of rabbits for sale in the marketplace. Only after all the rabbits had been consumed did the citizens notice that the neighborhood cats had all disappeared. We understood then the literal meaning of the old Spanish maxim *dar gato por liebre*.[6]

"…I go to the *malecón*," one suffering citizen in Havana wrote,[7] "to sell in dollars to the *jineteras* the clothes I no longer use, or to barter sugar for malanga, malanga for beans, beans for onions, onions for rice, rice for powdered milk, powdered milk for detergent, detergent for aspirin, aspirin for sugar, and so on, and so on, and so on…"

We were all terribly depressed after the visit to the patients' homes. The extent of the scarcity was beyond what we had imagined. Hunger was affecting the entire population of the island. Many people went without food for days because the rations were insufficient for an entire month. Nobody was dying of hunger, but everybody was hungry.

It was obvious that the economy of the island was in shambles. Cuba had been pulled to the bottom in the wreck of the Soviet Union and other communist countries. The tightening of the 30-year-old American economic embargo by the Torricelli law had guaranteed that the Cubans could find no new trading partners. Regulations that went into effect in mid-1992 made it illegal for any American company or any of its subsidiaries to conduct any business with Cuba and closed US ports to any vessel that had anchored in any Cuban port within the previous 180 days, and to any ship containing any Cuban product or even anything that *included* any Cuban product—steel, containing Cuban nickel, or fruit juice, containing Cuban sugar.

The embargo had become a blockade.[8]

The US embargo provided the Cuban people with an easy explanation for their misery. "Castro's government is doing everything possible," they told us. "We blame our hunger on the US embargo."

★★★

At the end of the long and depressing day, an unexpected tropical storm stranded us under the roof of a local school for nurses. We had a welcome opportunity for informal conversation with a group of nursing school students who had also taken shelter there.

These were young Cuban women, petite in size, many in their teens, some with blond hair and skin the color of the soft tan of molasses; some with dark brown hair, deep black eyes, and olive complexion; a few with dark skin and bright smiles. We asked them about the epidemic. Many of the nurses' relatives had been stricken.

"Do you take your vitamins?" Dr Caballero, the nutrition expert with the Argentinean accent, asked the group.

"No," one young girl answered. "They give you more appetite, and there is not enough food."

"What are you going to fix for supper tonight?"

"Rice and beans, of course!" they answered in a chorus.

"Do you go to sleep on an empty stomach?"

"Most of the time…"

"But, certainly, you go dancing during the weekend?" Venezuelan-born Dr Muci-Mendoza asked. "Cuba is famous for the rhythm of its dance music."

"Not anymore," one of the nurses replied. "With the blackouts, there is no music, no dances."

"Besides," another girl added, "*no hay ambiente*. Nobody feels like dancing."

Dr Caballero, who had spent all of his professional life studying nutrition, was visibly shaken. This was in blinding contrast to life in a country where food is plentiful, where every supermarket has rows of shelves dedicated to displaying only varieties of breakfast cereals, and where the main nutritional problem of the country is obesity. These young Cuban women with the healthy appetite of growing adolescents were surviving on one meal a day at the school cafeteria, burning a load of calories to ride their bikes for miles to get home to eat a plate of rice and beans, day in day out … The students told us they studied by candle light, in the heat of the tropical night, and then fell asleep and dreamed of food. Small wonder that the festive Cuban spirit was beginning to sink under the weight of the privations.

Date: Monday, 5/20/1993, Pinar del Río. As a medical student, during your training in Psychiatry, you are taught to remain impervious to the emotional pain of patients in order to be able to help them with therapy in a rational and level-headed manner. The training helps you to maintain, in most cases, a detached attitude; but, of course, you are only human, and it is often impossible to not empathize with the patients' grief, to feel their pain. This often is the cause of physicians' burnout when they carry too much of their patients' pain and misery on their shoulders.

In contrast, when you examine the unending number of victims of an epidemic, the individual patient loses his or her identity, and you begin to grieve with the pain and suffering of an entire nation as a way to express solidarity with the fate of the people.

All the frustration I had been experiencing as a member of the Mission to Cuba team, the sense of helplessness, and the disappointment in not being able to find the cause for the epidemic to help the victims and prevent future cases of blindness, were lifted today. A young country doctor gave us the answer—easy and simple: famine.

For this reason, I saw a profound injustice in this epidemic. There is hunger and misery everywhere. Today, I discovered also that politicians are the true culprits of this epidemic. It is ironic that both "epidemics" and "politics" are words based on the concept of *people*—epidemic from the Greek *epi-demos*, meaning harmful events descending upon people, and politics from the Latin *populus*, meaning actions undertaken for the people. But here in Cuba, people were being harmed by the political decisions of political leaders in Cuba and the United States. The Cuban leadership on the one side decided to follow a distant and failed social system and the US politicians on the other side decided to use the tools of democracy—another word based on the Greek *demos*—to force the surrender of a small nation that is not an enemy at war. In my view, it was clear that both political systems had simply forgotten that political decisions may have major repercussions on the health of the people. Politicians were acting like the blind guiding the blind. The advice I found in the Bible (*Matthew* 15:14) seemed disconcertingly relevant to the present situation: "Leave them alone. They are blind leaders of the blind. And if the blind guide the blind, both will fall into the same pit."

ENDNOTES

1. *Vapores* = hot steam vapor.
2. *Bodegas* = stores for rationed food distribution.

3. *Malanga* = cocoyam, also called *ñame*, a Cuban starchy-root staple similar to the Hawaiian taro root (Wikipedia).

4. *Caldoza* = (also spelled *caldosa*) Cuban potluck, Cuban stew.

5. In April 1993, Carlos Barros, Trade Commissioner of Cuba in Montreal, Canada, contacted Dr Lewis Rowland, Professor of Neurology at Columbia-Presbyterian in New York, who arranged for a visit of the first team of US physicians to the island to document the epidemic. On May 4–11, 1993, through the International Peace for Cuba Appeal, Drs Norah S. Lincoff and Jeffrey G. Odel from the Department of Ophthalmology, and neurologist Dr Michio Hirano, all from the College of Physicians and Surgeons of Columbia University in New York, visited Cuba and interviewed and examined patients with optic neuropathy. Their report, published as "A Letter from Havana" in *JAMA, the Journal of the American Medical Association*, became the first documentation in the medical literature on the Cuban epidemic. It provides a clear description of the deficient diets of these patients. See, Lincoff NS, JG Odel, M Hirano. "Outbreak" of peripheral neuropathy in Cuba? *JAMA* 1993; 270:511–518. During the first year of the epidemic the studies of the National Institutes of Epidemiology, Nutrition, and Toxicology noted that there were virtually no cases among children, teenagers, pregnant women, and the elderly. These groups received nutritional supplementation and were protected from the epidemic. Visual loss affected men preferentially, between the ages of 45 and 64 years, while burning feet were observed mainly in younger women of 25–44 years of age. The Cuban diet during the years preceding the epidemic was quite poor. A nutritionist in the group of Drs NS Lincoff, JG Odel, and M. Hirano (op. cit.) referred that a typical patient would eat mainly beans, potatoes, boiled vegetables, and non-enriched bread and rice; chicken once every 9 days and 1–2 eggs per week. The patient said that he never ate raw cassava, red meat, or milk. He had lost 12.6 kg during the previous 2 years (weight, 68.4 kg on admission). Another patient denied consuming any milk, chicken, or red meat in several months prior to the beginning of the disease.

6. *Dar gato por liebre* = literally, to give cat instead of hare or rabbit. To trick and rip off someone; to buy a pig in a poke; to have the wool pulled over the eyes; to be taken for a ride; a bait-and-switch scam.

7. Zoé Valdez. *La Nada Cotidiana* (Buenos Aires, 1996).

8. M. Murray. *Cruel and unusual punishment: The US blockade against Cuba* (Melbourne, 1993); Anthony F. Kirkpatrick, Richard Garfield and Wayne Smith. The time has come to lift the economic embargo against Cuba. *Journal of the Florida Medical Association* 1994;81:681–685; Wayne S. Smith. Cuba's Long Reform. *Foreign Affairs* 1996 (March/April); 75 (2): 99–112; L. Eisenberg. The sleep of reason produces monsters. Human costs of economic sanctions. *New England Journal of Medicine* 1997; 336: 1248–1250 (1997); Gustavo C. Román. Epidemic neuropathy in Cuba: A public health problem related to the Cuban Democracy Act of the United States. *Neuroepidemiology* 1998; 17: 111–115. For the controversy on the pernicious health effects of the embargo, see, *Annals of Internal Medicine*: http://www.annals.org/issues/current/full/200008150-00018.html#top.

CHAPTER 17

Cassava and Cyanide
Thursday, May 20, 1993

We thank the almighty God, for giving us cassava.
Cassava Song, Flora Nwapa, Nigeria

At the beginning of our 5th day on the island, we had no suspect to apprehend as a culprit for the epidemic. By now more than 6000 patients had been diagnosed since the Pan American Health Organization first began assembling the team and as many as 500 new patients had been admitted daily to hospitals on the island for treatment. We all had hoped that some answers could be found in the epicenter of the epidemic, the Province of Pinar del Río. That morning, after the conclusion of formal presentations at the auditorium, Dr Mariluz Rodríguez Alvarez—the blond, energetic, fast-talking epidemiologist, who knew all the data by heart—approached Dr Llanos and me while we were drinking our *café Cubano*. She inhaled the smoke from her black-tobacco "Popular" cigarette and then told us in a low confidential tone:

"You have probably noticed that increased consumption of *yuca* was statistically significant in the months preceding the onset of the epidemic. Do you think that this is something we need to look at in more detail?"

"Indeed," I replied, "in tropical neurology this is an alarm signal because cassava has been linked in Africa to epidemic diseases that affect sight and hearing."

Dr Llanos agreed to review the database with her team in more detail, while I went with the clinical team to visit the patients at their homes.

During the year preceding the epidemic, cultivation of cassava (*Manihot esculenta*) had been promoted in the province, and per-capita consumption of this starchy tuberous root had tripled during the first months of the epidemic.

★★★

Cassava—known as *yuca* in Spanish, *manioc* in French, and *mandioca* in Portuguese—is one of the world's most important food staples. In the tropics it is consumed daily by as many as 400 million people. The starch from cassava is known in different cultures as *tapioca*, *farofa*, and *casabe*. Fermented cassava is consumed as a gruel known in Africa as *gari* or *luku*.

Michele de Cunio, an Italian who traveled with Christopher Columbus on his second voyage to the New World, described in detail the preparation of a bread called *casaba*.[1] "(*Casaba*) may last in good condition as long as 15–20 days and often saved us from trouble." The thin *casabe* bread soon became a regular item of the ships' supplies and the staple food of the Conquistadors' provisions for the conquest of the Americas.

Cassava was a generous gift of the New World to Africa and to the tropical Old World—the equivalent of potatoes and maize that later saved Europe from starving. The Portuguese, probably through the delta of the River Niger, introduced cassava from Brazil into Africa. John Barbot, an explorer of West Africa in the late 1600s, wrote: "Magnoc bushes, which they call *Mandi-hoka* in their language; of which they make cassava or *Farinha de Pao*, that is, in Portuguese, wood-meal, which is the bread they commonly feed on."[2]

However, native Americans knew from their traditional wisdom that cassava must be carefully used because of its high content of prussic or hydrocyanic acid. The Spanish Conquistadors fell victim to one of the earliest documented episodes of cassava poisoning when the native Indians offered them fresh juice from the bitter cassava peels—resulting in acute, fatal cyanide poisoning.

Despite the difficulties in preparing it for consumption, cassava is a generous staple. Highly resistant to drought, it grows well in poor soils and has a potential yield of 75 tons/ha—twice as much as wheat. The plant is resistant to insects and pests because of the high content of cyanide-releasing substances or cyanogens found in the starch, the peel of the tubers, and the leaves.[3] In Africa, during long droughts, the only green plants in the scorched fields are the dark-green palmated leaves of the tall cassava plants. It is at these times that cassava becomes the only edible root crop available. When it is gulped down uncooked or when improperly detoxified by too short a period of fermentation or by sun drying, it produces a number of neurologic diseases.

Since the 1930s, British physicians had noted that malnourished Nigerians living in poor communities and surviving on bitter cassava as the staple food suffered from a condition that caused difficulty walking and loss of sensation in the legs due to damage of the spinal cord and the nerves. Dr Benjamin Osuntokun, an African neurologist from the University of Ibadan, called this disease "Nigerian tropical ataxic neuropathy" (TAN). Dr Osuntokun would finally discover the mysterious cause of this disease.[4] Dr Osuntokun, a Nigerian-born Yoruban, who became one of the most distinguished African physicians, conducted single-handedly, between 1966 and 1970, one of the most comprehensive studies ever undertaken in tropical neurology.

In a first group of 375 patients suffering from TAN, the dietary history confirmed in every instance the almost total dependence on a diet of bitter cassava with occasional supplements of yam, maize, rice, vegetables, and rarely animal protein. Most, however, "ate not other major item of food." All age groups from 10 years onward were affected. The disease began with a pins-and-needles sensation and burning pain in the feet, followed by blurring and loss of vision, trouble walking, noise in the ears leading to deafness, and then weakness and thinning of the legs.

Not unexpectedly, these patients had no detectable levels of sulfur-containing amino acids and had low riboflavin and vitamin B_{12} levels. Dr Osuntokun then concluded that "in Nigerians who suffer from TAN, chronic cyanide intoxication from cassava consumption appears to be the most important factor in the causation and pathogenesis of the disease." Dr Osuntokun's critics, however, considered that deficiency of thiamine (vitamin B_1), vitamin B_{12}, and riboflavin (vitamin B_2) were the most important factors.

New evidence of cassava toxicity would be provided 20 years after Dr Osuntokun's work by a study on *konzo*, which was undertaken by two researchers from the International Health Unit of the Department of Pediatrics of Uppsala University in Sweden.[5]

Dr Hans Rosling and Dr Thorkild Tylleskär dedicated a large part of their scientific careers to the study of cassava and *konzo*, an exotic problem occurring a continent far away from the land of the midnight sun where they had grown up.[6]

The triad of symptoms described by Dr Osuntokun in hundreds of his patients was very similar to that of the Cuban patients: loss of vision, burning feet, and deafness. There was also documented evidence that cassava consumption had increased in Pinar del Río during the *periodo especial* just prior to the beginning of the epidemic. Therefore, there was a good possibility that cassava could be the culprit causing the Cuban epidemic.

***

While Dr Llanos was investigating the possibility that the epidemic of blindness might be linked to cassava toxins, other members of the Mission team were stalking another culprit. In our conversations with the people of Pinar del Río, we had learned about *caña santa*, or lemon grass (*Cymbopogon citratus*), which the impoverished citizenry was using as a substitute dessert or stimulant. Coffee, tea, chocolate, the Cubans' beloved *Copelia* ice cream, and *Tropicola* sodas had all but disappeared from the shops and stores. So people began substituting them with a drink made from an infusion of a small grass with a lemony taste, sweetened with sugar. In workplaces, in schools, and at home, *caña santa* became Cuba's most popular drink. Similar popular infusions were being made from orange tree leaves, mint (*Mentha piperita*), and chamomile (*Matricaria chamomilla*).

Dr Caballero and I had our suspicions. Neurotoxicity has been suspected from frequent use in the tropics of teas made from soursop leaves (*Annona muricata*). We knew we needed to learn more about these plant infusions.

Now, we had at least two possible suspects—cassava and lemon grass— from the vegetable kingdom. Also, there was the possibility of cyanide toxicity from beans, plantains, and even sugar cane. Consultation with experts in plant toxicology would be required to answer these questions.

Before we could consult with them, though, more possible culprits were being identified. A key one was methanol.[7]

Throughout modern history, a frequent cause of outbreaks of blindness in the tropics is the adulteration of liquors with methanol. Methyl alcohol, or wood alcohol, is produced by fermentation of bagasse—the dry pulp that remains after the juice of the sugar cane has been extracted. Pure methanol is transparent, colorless, odorless, and tasteless. It is also inexpensive, and for this reason it is mixed with gasoline to produce the automobile fuel called "gasohol."

For the same reasons it is also an ideal adulterant of expensive liquors, such as whisky, vodka, or even aged rum. But methanol is dangerous, and methanol intoxication can produce blindness or death. In an attempt to eliminate the alcohol, the body transforms methanol into formic acid, which results in destruction of the optic nerves and, in severe cases, a lethal acidosis. In Vermont, where methanol is used as antifreeze, the poison is also a drug of choice for the surge of suicides that occurs toward the end of the long winters, the so-called cabin fever.

Dr Sadun suggested that due to the extensive practice of producing homemade alcoholic beverages in Cuba during the *período especial*, methanol could have contributed to the loss of vision.[8] Many Cubans, it was known, were making their own alcohol at home, producing it by fermenting fruit juices, or sugar water, until it became an intoxicating beverage. Could some of them have been using methanol, too?

There was perhaps a fourth culprit still to be identified: Strachan disease.[9]

All members of the Mission had read the description of an epidemic of blindness described by Dr Henry Strachan in Jamaica, then a British colony, in 1888, the same year of the infamous crimes of "Jack the Ripper" in London. In both cases the culprits had never been identified.

Dr Strachan was a British physician, who served as Senior Medical Officer at the Kingston Public Hospital in Jamaica. He reported the occurrence of an outbreak of more than 500 cases of blindness, deafness, burning feet, and paralysis. Dr Strachan's patients were black peasants, young adult men and women working in the sugar cane plantations; they were afflicted by the following condition:

> A patient presents himself complaining of numbness and cramps in his hands and feet, dimness of sight, and a tightness round the waist. He may add to this statement that he fears he is getting hard of hearing. He goes on to say that he suffers from severe burning in the palms of the hands and soles of the feet (…)
>
> On admitting such a case to hospital, and watching its further progress, it will be noted that at night the patient will be awake for hours, rubbing his feet and legs, and moaning with pain. The loss of vision will proceed until he cannot recognize individuals. The muscles of his limbs will waste until the "claw" hand and foot are marked features (…) In the feet, the thick epidermis of the soles of such patients as walk barefoot will come off in large flakes, and the soles will become… deeply pigmented as the disease progresses, and remain so after it has passed away.

The cause of this outbreak remains unknown to date, although most experts consider that Strachan disease was caused by a poorly defined nutritional deficiency. This tropical neurologic mystery had proven irresistible to the epidemiologist–medical detective in me. In 1985, I wrote in the *Transactions & Studies of the College of Physicians of Philadelphia* a historical review on Strachan disease.[10]

In the best detective style, I proposed that the cause of this mysterious disease had been arsenic poisoning. The reason for postulating arsenic as the causative factor was the discovery of an epidemic outbreak of arsenic intoxication in England, contemporaneous with the Jamaican cluster. During my historical research I found a report by Dr Ernest S.

Reynolds from the Workhouse Infirmary in Manchester, England, titled "An account of the epidemic outbreak of arsenical poisoning occurring in beer-drinkers in the north of England and the midland counties in 1900." This epidemic produced 6000 cases of an illness which manifested as painful numbness of feet and hands, blindness, hearing loss, desquamation of the soles and palms, and a typical discoloration of the skin that gave the patients the appearance of being sun-tanned (a most unlikely event among beer-drinkers in Manchester in the 1900s). Dr Reynolds then discovered that the source of arsenic was "the whitest sugar" imported from the West Indies and used in the brewing of beer.

Therefore, I had concluded that both epidemics—Strachan disease in Jamaica in 1888 and Reynolds disease in England in 1900—could have been caused by arsenic from a common source of poisoning in sugar cane plantations in Jamaica and the British West Indies. Clinically, the combination of skin pigmentation, blindness, hearing loss, burning feet, and wasting of the muscles of hands and feet described by Strachan is identical to the clinical picture described by Reynolds in thousands of his patients and therefore typical of arsenic intoxication. The source of the arsenic in the beer was the sugar, which, in turn, had been contaminated by arsenic compounds—including the popular Paris green (copper acetoarsenite)—commonly used as weed killers and pesticides in sugar cane fields.

Was it possible that now, almost 100 years later, arsenic had again contaminated the sugar being consumed—in increasingly larger amounts, to substitute for more traditional foodstuffs—by the impoverished Cubans?

★★★

The return trip from Pinar del Río was made quite short by the animated discussions among the members of the Mission on the findings of the day. We were all clearly impacted by the severity of the lack of food and the poverty of the region, but none of us—including Dr Caballero, the team's nutrition expert—had had any previous experience with nutritional deficits capable of producing outbreaks of neurologic disease of the magnitude and severity of what we had witnessed today. We all knew of examples from many wars throughout human history that vitamin deficiencies as a result of poor diets are, indeed, a well-known cause of neurologic diseases. But we had never encountered this directly, and we were all reluctant to commit ourselves to a single nutritional cause of the epidemic. Moreover, typical groups, such as children, pregnant and lactating women, and the elderly, who are usually affected first and more severely in cases of nutritional deficits had been spared from the epidemic in Cuba. More detailed information on the intricacies of the food rationing in Cuba would be needed to better understand the behavior of the epidemic.

Perhaps for those reasons, we embraced enthusiastically the four neurotoxic hypotheses that had been raised during the day. Regarding cyanide toxicity from cassava, Dr Llanos confirmed that in response to an initiative led by Fidel Castro, cultivation and consumption of *yuca* in the region had been promoted and the exhaustive review of Dr Mariluz Rodríguez's databases provided the figures corroborating that consumption of *yuca* had tripled in the Province months prior to the diagnosis of the first cases of blindness. These figures were available by week and by municipalities in the entire Province, but Doctor Llanos noted one striking discrepancy: a dose-effect was not present. Therefore, municipalities with the highest *yuca* consumption that should have had the largest number of cases did not, whereas many cases had occurred in places with only modest increases in *yuca* consumption above base line figures.

"We also need to know if the consumption of *yuca* increased on the island along with the epidemic," Dr Muci-Mendoza observed.

"And there is also the well-known fact of cassava toxicity," I added. "Neurologic diseases have only been reported in Africa, where bitter varieties of cassava with high-cyanide content are cultivated. To my knowledge, there are no outbreaks reported from the Americas or India, despite similar rates of cassava consumption. We will need an expert to tell us if high-cyanide bitter cassava is cultivated in Cuba."

From cyanide in cassava we moved next to toxic herbal teas. Dr Caballero reminded us that during times of nutritional deprivation, with unavailability of the usual foodstuffs, people resort to consuming any available products, including those that are toxic in nature. One of the best known cases involved the consumption, during the Spanish Civil War, of the hardy and toxic *Lathyrus sativus*, a wild variety of sweet pea, which could cause permanent paralysis of the legs. In Cuba, serious consideration had to be given to this toxic possibility because of the increased consumption of herbal teas made from infusions of *caña santa* and a number of other herbal products. Again, this was another case to discuss with a toxicology consultant.

With less enthusiasm the group discussed the distant third possibility, Professor Sadun's hypothesis of methanol intoxication causing a mitochondrial optic nerve energy deficiency compounded by the lack of nutrients that normally help detoxify methanol. There were several arguments against this theory: the clinical picture of a blinding hangover was not found here; instead, in Cuba, the patients' visual loss developed in a slowly progressive manner over weeks and months. More importantly, patients who denied drinking any alcoholic beverages had also developed the blindness. Lastly, methanol intoxication often causes fatal acidosis, but increases in such fatalities had not been reported. It appeared to us that we could tentatively table further discussion of the methanol hypothesis; however, we would certainly ask the toxicology consultant about this neurotoxin as well. For the moment, the Mission team concluded that future specific research would be required to demonstrate formic acid in samples of spinal fluid from Cuban patients with epidemic optic neuropathy in order to demonstrate the feasibility of Professor Sadun's hypothesis.

And last but not least, there was the disease of still undetermined cause described by Dr Strachan in Jamaica. We knew that our Cuban neuroophthalmology colleagues—in particular Dr Rosaralis Santiesteban—favored this diagnosis. It was evident at once that this historical precedent shared many characteristics of the current epidemic. The tropical Caribbean environment of Cuba and Jamaica were identical, and the physical manifestations, such as dimness of eyesight, hearing loss, and painful sensations of palms of hands and soles of feet, were suspiciously similar. However, there were also significant differences: the wasting of the hand muscles resulting in a claw-hand, as described by Strachan, did not occur in the Cuban patients, nor did the hard-to-miss changes in the skin and its pigmentation. Finally, Dr Strachan had believed that this was a form of malarial neuritis, but malaria—along with yellow fever—had been eradicated in Cuba years ago.

"Do not forget," I told the team, "that the symptoms of Strachan disease are consistent with arsenic intoxication; and remember also that arsenic has been, for years, a popular herbicide and pesticide in sugar cane plantations."

It was now clear that interrogation of these four suspects—cyanide from cassava; *caña santa* and other herbal teas; methanol in home-fermented alcoholic drinks; and arsenic intoxication in Strachan disease— would require the expertise of a genuine neurotoxicology expert.

ENDNOTES

1. Michel de Cúneo, Letter to Jerónimo Annari, Savona, October 15, 1495. Quoted by Alberto M. Salas and Andrés R. Vásquez. *Noticias de la tierra nueva* (Buenos Aires, 1964). Reprinted in *Cronistas de Indias, Antología* (Bogotá, 1982). The pertinent text reads (p 30): "These islands produce large roots like radishes (meaning cassava) very white, and large; they produce bread as follows: the roots are grated like we do with cheese, using a rough stone, they have in the fire a large stone and on its surface they place the grated root producing a sort of flat bread that lasts in good condition 15–20 days and that many times was our only sustenance."

2. E.J. Alagoa. Long-distance trade and states in the Niger Delta. *Journal of African History* 1970; 11 (3): 319–329.

3. Barry Nestel, Reginald MacIntire (editors). *Chronic Cassava Toxicity* (IDRC, Ottawa, Canada 1973); AM Ermans, NM Mbulamoko, F Delange, R Ahluwalia (editors). *Role of Cassava in the Etiology of Endemic Goitre and Cretinism* (IDRC, Ottawa, Canada 1980); M Bokanga, AJA Essers, N Poulter, H Rosling, O Tewe (editors). *International Workshop on Cassava Safety. Acta Horticulturae* Number 375 (November 1994); Hans Rosling. *Cassava Toxicity and Food Security* (Uppsala, 1987); Nicholas LV Mlingi. Cassava Processing and Dietary Exposure in Tanzania. *Acta Universitatis Upsaliensis* 571.69 pp (1995).

4. Benjamin Osuntokun. An ataxic neuropathy in Nigeria: A clinical, biochemical and electrophysiologic study. *Brain* 1968; 91:215–248; BO Osuntokun, et al. Tropical amblyopia in Nigerians. *Am J Ophthalmol* 1971;71:708–716; BO Osuntokun, et al. Epidemiology of tropical nutritional neuropathy in Nigerians. *Trans R Soc Trop Med Hyg* 65:454–479 (1971); BO Osuntokun, et al. Cassava diet, chronic cyanide intoxication and neuropathy in Nigerian Africans. *World Rev Nutr Diet* 36:1412–173 (1981); Gustavo C Román. Obituary: Benjamin Oluwakayode Osuntokun. *J Neurol Sci* 147:1–3 (1997).

5. *Konzo* = a neurologic disease of poor African children fed cassava exclusively.

6. Hans Rosling, Cassava, cyanide, and epidemic spastic paraparesis. *Acta Universitatis Upsaliensis* 1986;19:1–52 (1986); H Rosling. *Cassava toxicity and food security* (Uppsala, 1987); Thorkild Tylleskär. The causation of konzo. Studies on a paralytic disease in Africa. *Acta Universitatis Upsaliensis* 1994;43:1–67 (1994); G. Trolly, "Konzo:" Epidemic Spastic Paraplegia of the Congo Natives, according to the physicians of the Queen Elizabeth Foundation for Medical Assistance to the Belgian Congo Natives. Brussels: 112 rue du Commerce, 1938. (Abstract in, *Tropical Diseases Bulletin*, vol 36, No. 6, June 1939:501–502); H Carton et al. Epidemic spastic paraparesis in Bandundu (Zaire). *J Neurol Neurosurg Psychiatry* 1986;49:620–627; Ministry of Health, Mozambique. Mantakassa. *Bull WHO* 1984;62:477–484, 485–492; J Cliff et al, Association of high cyanide and low sulphur intake in cassava-induced spastic paraparesis. *Lancet* 1985; ii:1211–1213.

7. CD Benton, Calhoun FP: The ocular effects of methyl alcohol poisoning. Report of a catastrophe involving three hundred and twenty persons. *Trans Am Acad Ophthalmol* 1952, 56:875–883; Janis T Eells, Methanol. In, *Browning's Toxicity and Metabolism of Industrial Solvents*, 2nd ed, Vol 3: Alcohols and Esters. Thurman RG, Kauffman FC (eds), (Amsterdam, 1992) pp 3–15; ——, Salzman MM, Trusk TC, Inhibition of retinal mitochondrial function in methanol intoxication. *Toxicologist* 1995, 15:21; ——, Methanol-induced visual toxicity in the rat. *J Pharmacol Exper Therap* 1991, 257:56–63; Murray TG, Burton TC, Rajani C, Lewandowski MF, Burke JM, Eells JT: Methanol poisoning: A rodent model with structural and functional evidence for retinal involvement. *Arch Ophthalmol* 1991, 109:1012–1016; ——, Salzman MM, Lewandowski MF, Murray TG, Development and characterization of a nonprimate animal model of methanol-induced neurotoxicity. *Environmental Toxicology and Risk Assessment: Biomarkers and Risk Assessment* (5th vol.), ASTM STP 1306, Bengston DA, Henshel DS (eds), American Society for Testing Materials, Philadelphia, 1996, pp 239–254.

8. Janis T Eells, González-Quevedo A, McMartin KE, Sadun AA. Folate-deficiency and elevated serum and CSF formate concentrations in patients with Cuban epidemic optic neuropathy. *Investigations in Ophthalmology and Visual Sciences* 1996,37:S496.

9. Henry Strachan, Malarial multiple peripheral neuritis prevalent in the West Indies. *Sajou's Annual of the Universal Medical Sciences* 1:139–141 (1888); ——, On a form of multiple neuritis prevalent in the West Indies. *Practitioner* 59:477–484 (1897). Strachan's syndrome is the name frequently used for the combination of visual loss, hearing deficits, and neuropathy in the tropics, regardless of its cause. His description continues as follows:

On examination it will be seen that there is slight excoriation of the edges of the eyelids, margins of the lips, and around the margins of the nostrils; the palpebral conjunctive may be hyperemic, as will be the lips. The heat of the hands complained by the patient will be found to be not merely subjective but appreciable to the touch, and due to a hyperemic condition of the palms; the acuteness of vision for form will be found to be more or less impaired according to the stage to which the malady has progressed; examination of the main nerves to the extremities will show that they are very tender on pressure, especially the ulnar nerve, and along the distribution of their terminal filaments they may be tracked by lines of fine herpetic vesicles.

On admitting such a case to hospital, and watching its further progress, it will be noted that at night the patient will be awake for hours, rubbing his feet and legs, most probably, and moaning with pain. The loss of vision will proceed until he cannot recognize individuals. The muscles of his limbs will waste until the "claw" hand and foot are marked features, and the wasting of muscles and the disappearance of fat will produce emaciation which is very noteworthy in advanced cases. There will be found to be no alteration in the reaction of the pupil to light and accommodation, no falling when the eyes are closed, and the sphincters will not be affected. All patients walked with a typically ataxic gait, and in some there was also involvement of the upper extremities, any attempts to move resulting in a peculiar aimless jerk. The knee jerks were absent in more than half of the patients (53%), exaggerated or normal in 23%, and normal in the rest. Pain, touch and temperature sensations, although delayed, were completely absent only in the more severe cases. Retinal hyperemia was noted, but there was no optic atrophy.

This desquamation, which is fine and branny, involves the whole of the palm, which becomes more and more deeply pigmented as the disease progresses. Should bullae form and be neglected, small ulcers may result, but this is apparently rare. In the feet the thick epidermis of the soles of such patients as walk barefoot will come off in large flakes, and the soles also will become pigmented…These parts, which are normally the least pigmented external portions of the negroes, become deeply pigmented as the disease progresses, and remain so after it has passed away….The change is most striking to the observer in the regions, which are normally fairer than the rest. Then the color may vary from brown to intense black.

In a few fatal cases resulting from respiratory distress, autopsies showed dark pigmentation of the brain, spinal cord and nerve trunks. Liver and spleen showed the tell-tale signs of chronic malaria.

The latter findings led Strachan to believe that malarial pigment was the cause of the disease he called, "Malarial Multiple Peripheral Neuritis."

10. Gustavo Román. Epidemic neuropathies of Jamaica. *Transactions & Studies of the College of Physicians of Philadelphia* 1985; series 5, no 4, 7:261–274. The possible etiology of Strachan's syndrome as arsenic poisoning is explored in this historical review. See, Ernest S Reynolds. An account of the epidemic out-break of arsenical poisoning occurring in beer-drinkers in the north of England and the midland counties in 1900. *Lancet* 1901;i:166–170. This epidemic of 6000 cases was caused by "the whitest sugar" from the West Indies, contaminated with arsenic and used in beer produced in the North of England.

CHAPTER 18

The Myelitis of Santiago de Cuba

The Pharaoh's right leg with pronounced hypotrophy of the muscles and with an equine foot is typical of poliomyelitis.
Bas-Relief of Pharaoh of the XVIIIth Dynasty (1500 BC)

Date: Friday night, 5/21/1993, Havana, Cuba. The days had passed swiftly, filled with a sense of urgency and excitement. The consequences of the embargo and the *período especial* were obvious everywhere: instead of buses, public transportation relied on long trucks lined with cattle fences where people travel crowded with standing room only. People walked or rode bicycles, and you could see on the streets two-wheel carts pulled by mules. There were long periods of blackouts, with no electricity for refrigeration or air conditioning.

The next day would be the end of our first week in Cuba. In these few days on the island, we had made limited but solid progress: we could now clearly identify the cases of optic neuropathy, and we also had defined the peripheral manifestations of the Cuban epidemic neuropathy. We had seen enough patients to recognize the common and uncommon manifestations of this epidemic disease and we felt that we had covered the first half of the first mandate of our Mission: "Review and analyze available information on the clinical and epidemiologic aspects of the epidemic."

With regard to the second half of our first task "identify possible causative factors of the ongoing epidemic of optic neuritis in Cuba," we had proposed four potential toxins as culprits. But the Cuban Chief of Toxicology, Dr Pérez-Cristiá, assured us that the tests for cyanide, arsenic, and other heavy metals had been completely negative in all patients affected with epidemic neuropathy. Therefore, the first suspects of the Cuban epidemic—cyanide from cassava and arsenic in Strachan disease—had been lined up and released.

The nightly briefings had been made more dramatic by Fidel Castro's appearances. Soon after sunset, every night—but never at the same hour—we observed the speeding jeeps of the presidential guard arriving under the canopy of the Biocaribe Hotel. Armed soldiers, in a state of full alert, spilled from the jeeps and took positions on the hotel steps.

Next, immediately behind the guards, two black armor-plated Mercedes Benz limousines arrived, followed closely by another jeep carrying armed soldiers in fatigues. From one of the limousines first Dr Selam Houssein, Castro's personal physician, emerged, and then after a brief pause, *el Comandante* himself appeared. Preceded by his personal bodyguards dressed in *guayabera* shirts and surrounded by soldiers, he was dressed always in his traditional dark green fatigues and trademark military cap, and the aging leader would slowly climb the front steps to the hotel.

Castro then joined the Mission members in the conference room where the scientists would share their findings, ideas, and theories. The *Comandante* would greet each member of the Mission with a handshake and make brief comments specific to each one of us. Then he would take his place at the podium at the head of the assembly, where microphones, lights, simultaneous translation, and recording equipment would already be positioned. He would open the conference by asking for a summary of the activities of the day and the plans for the immediate future. Throughout the visit, Castro would demonstrate perfect familiarity with the topic at hand, for instance, quoting with precision data on the number of cases of the epidemic at particular times or in a given city, or the types of treatment being used, or the superior results obtained with methylcobalamin versus cyanocobalamin, and the dosages used.

It was obvious that no other issue in Cuba was more important to Castro than the epidemic.

The Mission's virologist Dr Gajdusek wrote in his diary:

> ...*20 May 1993...Hotel Biocaribe, Havana, Cuba*
> *I did not anticipate in coming to Cuba that I would be talking for hours with Fidel Castro. As I have heard repeatedly from others, he has a piercing intelligence, which quickly grasps all that transpires around him and he has enormous ability to quickly précis complex data into succinct, glib and accurate summaries. He is no Dionysian, but rather an Apollonian ascetic, clearly living for his ideal. He must be about my age. He dresses in his military greens and mingles closely with the crowd even at our banquet and reception in an alert, relaxed fashion. Only when he leaves and quickly is surrounded by his bodyguards does one perceive that he is a man of power always in danger. This shows through also at the sessions he chairs where, with his precise summaries, he always demands to know just where we stand and what the next step is.*[1]

During the evening session on May 19, midway through our first week in Havana, we had been informed of the recent occurrence of an unusual number of cases with symptoms of spinal cord involvement in Santiago de

Cuba, the most important city of the old Eastern Province, la *Provincia de Oriente.*

Patients observed there had become unable to walk, had lost control of bladder and bowel functions, and the affected men had become impotent. Evaluation of such patients in April 1993 by the current Director of Cuba's Neurological Institute Dr Ricardo Santiago-Luis González and neurologist Dr Rafael Estrada Jr, son of the Institute's first director, indicated that the clinical manifestations appeared to be more extensive and more severe than the isolated blindness that affected most patients, indicating spinal cord involvement.[2]

Their clinical diagnosis was "funicular spinal cord degeneration from vitamin B_{12} deficiency."

As often occurs, new names were coined for old diseases. The Cubans who first suffered this illness called it "the myelitis of Santiago de Cuba" and believed that it was a contagious disease completely different from optic neuropathy. Rumors had increased since news media had announced that a virus had been isolated from patients with neuropathy and even more with the arrival of the virologists from the United States. In the street the comment was: "Something serious must be happening for the Americans to send over their Nobel Prize scientist to study it."

Myelitis simply means "inflammation of the spinal cord." The best known form of myelitis is polio—the infection of the spinal cord by the poliovirus causing paralytic poliomyelitis. Likewise, *encephalitis* means "inflammation of the brain," often caused by a virus. Both terms and diagnoses put terror in the minds of doctors and patients alike because of the devastation that can be caused by the infections in the brain and the spinal cord despite all efforts to treat them. Nobody in Cuba wanted to become infected with this virus, the myelitis of Santiago, considered "the worst form of the disease." Because of the seriousness of this disease, it was determined that some members of the mission team should visit Santiago at once.

Early the following morning, the Mission team members were taken to the military airport of San Antonio de los Baños at the outskirts of Havana for a trip to Santiago de Cuba. Cuban Air Force personnel wearing dark purple berets saluted the team as the Mission members passed through the gate. A modern executive Cubana Airlines jet, originally a Soviet YAK-40, waited by the runway. The interior of the plane was designed like a meeting room with a central round table, comfortable chairs, and a sofa. The signs for seat belts and no-smoking were in Cyrillic.

Earlier in May this year, Castro had visited three hospitals in Santiago de Cuba—the Clinical-and-Surgical Hospital, the Saturnino Lora Hospital, and the Ambrosio Grillo Hospital. Castro had personally greeted about 200 neuropathy patients, shaking hands with them and even embracing an old *compañero* from the days of the Sierra Maestra. Because it was Mothers' Day, he had greeted some of the female patients with kisses on both cheeks in the Spanish manner.[3]

In this quiet but eloquent way he reassured his people that the neuropathy was *not* contagious. President Castro probably kept to himself the images of the virus isolated in Havana. Chillingly, the first viral isolations had, indeed, been obtained from patients hospitalized in Santiago de Cuba.

Our flight was brief, and we soon landed in the city of Santiago de Cuba. Deep inside a natural bay protected by the Morro Castle, Santiago extends uphill on the first elevations of the Sierra Maestra mountains. The city maintains a colonial Spanish flavor, with cobblestone streets and red-tiles houses ornamented by elaborate wooden balconies. The Mission members were packed into the now-familiar Russian Lada sedan cars, driven through the charming old town, and taken to the Saturnino Lora Hospital to examine patients with myelitis.

In case after case the presentation was the same. The patients were dressed in clean clothes that were almost in rags, and they were all clearly malnourished and incontinent of urine and stools. They walked slowly, dragging their feet as if fighting an invisible current, their legs having lost feeling in their feet and toes, but the knee reflexes were brisk. In contrast to the neuropathy patients elsewhere in Cuba, though, eye problems were severe only in rare cases. A few were experiencing hearing loss. None reported burning feet. The diagnosis was clearly a lesion of the spinal cord caused by lack of vitamin B_{12}. Most likely this was due to nutritional deficits. Was this myelitis part of the neuropathy, or were we facing a different disease?

After seeing the hospitalized patients, we were taken to visit some patients in their homes. As in Pinar del Río, again, we were confronted by the same poverty, the same empty pantries, and the same lack of food. We separated into small groups, and I went with a young neurologist colleague to visit the home assigned to us. I was struck by the severity of the disease in one particular patient.

A former dancer, Xiomara had kept the *nom-de-guerre* of her days as a chorus line girl at the Tropicana nightclub in Havana, famous for the beauty of its *mulatas*. In 1959, after the triumph of the revolution, the Tropicana was closed, and Xiomara married and had three daughters, who had then made her *abuela* many times. When we entered Xiomara's home, the strong ammonia stench of urine surprised and halted us in our tracks.

From her bed Xiomara apologized: "*Doctores*, good morning; forgive the smell, but I cannot control my bladder, and the urine is constantly leaking. My daughter keeps hand-washing the *paños*, the cotton rags, but it is difficult to stay dry."

My colleague answered politely, "Not to worry, *Señora*. We are doctors, and we are used to these problems."

I examined Xiomara meticulously. She was a black woman, lanky and very thin, and barely able to walk unsupported with stiff and uncertain steps. The muscles of the famously long legs of the chorus line *mulata* were now atrophied bare threads. She was scarcely able to lift her legs from the floor, move her toes, or pick up her feet. However, her knees and the legs demonstrated a steady resistance to the movements required on them. I forced the feet to move with a sudden push, and this resulted in *clonus*—a constant rhythmic jerking of feet like a fast-moving pendulum that finally stopped. When the knee was tapped once with the reflex hammer, the leg kicked briskly out of the bed, but tapping of the Achilles tendons elicited no response. I scratched the soles of the feet with a key, and the toes slowly separated, while the big toe moved up in an involuntary movement called the *Babinski sign*. Although Xiomara complained of numbness of the legs, she could still feel the sharp end of a safety pin in her thighs, but she had no perception of pain below her knees. Moreover, she could not feel the vibration of a tuning fork on her legs or arms and could only feel it on her forehead, collarbones, and elbows. I confirmed that she was incontinent of urine and stools.

Examination of the eyes with the ophthalmoscope revealed the mother-of-pearl pallor of dying optic nerves and confirmed what her daughter already knew, that Xiomara was blind. She could hardly see a light being shined into her eyes and could not count the fingers held in front of her face. Politely, my colleague asked Xiomara's daughter what foods her mother had eaten during the past 4 weeks. The answer confirmed our suspicions: only rice and beans, brown sugar, and occasionally potatoes. Once a week, the family received a soy-based *picadillo*, the Cuban equivalent of Hamburger Helper. Xiomara had lost her appetite, and some days she had only one meal, and some days no food at all. For months, the family had seen no dairy products, eggs, meat, fish, bread, pasta, fruit, or leafy green vegetables.

I told Xiomara and her daughter that she was suffering from the myelitis of Santiago; a spinal cord disease produced by lack of vitamin B_{12}, and that she should be treated in the hospital with intravenous vitamins, a balanced diet, and physical therapy. She also suffered from advanced Cuban blindness, the epidemic optic neuropathy. After conferring with her daughter, my colleague called for the ambulance to transfer Xiomara to the Saturnino Lora Hospital. She needed urgent treatment with cobalamin (vitamin B_{12}) and B-complex (vitamins B_1, B_2, and B_6) injections, oral supplements of vitamins A, E, and folic acid. More importantly, she was to be placed in an immediate program of nutritional recovery with a high-protein diet administered for a period of at least 10 days. She would also benefit from a permanent catheter in the bladder to control the skin burns caused by the leaking urine and the constant humidity, heralding the development of terrible, often fatal decubitus ulcers.

We then went back to our colleagues, and the entire Mission team went to visit two patients from whom positive viral isolations had been obtained. The first patient was a thin young man, who had recovered completely from the neuropathy after being treated with intravenous vitamins and a special diet but who appeared to have a chronic "red eye," a severe conjunctivitis.

The home of the second patient was located on a narrow street that went up the hill. The young woman with gypsy eyes had been suffering from frequent vomiting during her early pregnancy. She had lost her appetite, eventually suffered a miscarriage, and was still feeling weak and depressed.

She worked as an ophthalmology technician and had correctly diagnosed herself as suffering from optic neuropathy. In the hospital she underwent a spinal tap for cerebrospinal fluid analysis despite the fact that she had no evidence of involvement of the peripheral nerves or the spinal cord. From this test it was determined that she was suffering from the virus that had been isolated in Havana. Following a few days of hospitalization, treatment with intravenous vitamins, and a nutritious diet, she had recovered almost completely.

Despite food restrictions, the laws of hospitality had to be respected. Our visit was marked by neighbors bringing freshly made coffee. The tiny home filled with the famous aroma of the *café de Santiago*, from the nearby Sierra Maestra.

Small coffee cups were offered to the 20 or so persons in our group. Dr Asher glanced toward Dr Caballero, and Dr Caballero looked at me—all of us thinking the same thing: was it safe to drink from these cups in the home of the patient with a positive viral isolation? We were not sure, but we did not want to break the unwritten laws of civility. We accepted the coffee because, at least temporarily, the importance of the gesture of hospitality and human feeling took precedence over our concerns about our own health and the results of scientific research.

Outside, life went on as usual. Mule-drawn carriages from another time filled the streets with the rhythmic, metallic noise of horseshoes on cobblestones.

★★★

The Mission team returned to Havana that night to the usual evening debriefing with *Comandante* Castro. About 100 people filled the conference room: myelitis was making the news.

I was asked to summarize the neurologic findings. "The evidence indicating the occurrence of cases with spinal cord involvement is quite clear," I stated. "Perhaps, the only criticism is the use of an old name, funicular myelopathy. The current terminology is 'dorsolateral myelopathy' or 'subacute combined degeneration of the spinal cord.' Regardless of the name, this is a classic disease in neurology—one of the few emblematic lesions of the nervous system produced by nutritional deficiency. I am certain that the spinal cord damage in these patients is due to lack of cobalamin, or vitamin B_{12}."

In this way, the Mission team agreed completely with neurologists Dr Ricardo Santiago-Luis and Dr Rafael Estrada, Jr, from the Neurological Institute in Havana, who in April—just ahead of Castro's visit—had recommended urgent treatment of these patients with vitamin B_{12} injections.

Dr Gajdusek had not traveled with us to Santiago. Instead, the virologist had spent most of his day visiting the Tropical Medicine Institute and the laboratories of Biotechnology. He spent the afternoon examining patients at the Hermanos Ameijeiras Hospital located in the imposing 20-story building of the old National Bank of Cuba on the *Malecón*, the sweeping waterfront boulevard invariably featured on the picture postcards of Havana.

He was most comfortable talking to the virologist colleagues in the common language that creates a universal scientific brotherhood. He had made them feel at ease in his expansive friendly way, his easy smile, and the long anecdotes about Russian scientists—many of them familiar to his new Cuban friends because they had trained in their laboratories. But behind the relaxed demeanor, he was paying meticulous attention to all the technicalities being described and often asked the Cuban scientists to repeat a finding or to clarify a particular detail.

Dr Gajdusek was fully aware of the responsibility placed on his shoulders. Cubans were going blind at a rate of 500 new cases per day, and new cases of myelitis were being reported in the area of Santiago. If a virus was the causal agent of these diseases, then the entire future of Cuba's health was in his hands.

Discovering a virus that causes a new disease is like describing a new continent or a new constellation. The discovery is only the beginning, and the exploration can take decades. In this particular instance—should the discovery be confirmed—the Cuban scientists would inform the world that this virus was the agent of the epidemic disease, the one and only cause of the thousands of cases already reported and of the hundreds of blind people being added daily to the count. If confirmed, this virus would also require the immediate and total isolation of Cuba by an international quarantine to prevent the spread of the yet unknown infectious agent to other parts of the world. Cubans would be shunned like the lepers of the Old Testament; every Cuban would be *persona non grata* everywhere in the world. The country would not be capable of withstanding the combined effects of the current isolation by the US embargo and a possible quarantine. This new virus would mean the end of Cuba. Like the West African nations of Congo, Uganda, Liberia, Guinea, Gabon, Sierra Leone, and Sudan during the epidemic of Ebola two decades later, the country would be isolated forever.[4]

That night, following the report on our visit to Santiago de Cuba, Dr Gajdusek stood up to speak. For the next 40 min he presented compelling evidence *against* the possibility of this virus being the cause of the epidemic.

His arguments were simple but irrefutable: two of the laboratories had reported positive isolations in 80–90% of all cerebrospinal samples taken from patients suffering from myelitis. Dr Gajdusek had reviewed in detail the methodology used in both laboratories and had found it to be flawless. Both groups had large experience in viral isolations and had executed their tests without error.

He then interviewed and examined seven patients in whom the virus had been isolated. His findings were highly interesting. Two had optic neuropathy. Four had the peripheral form of the disease and complained of hot "burning vapor" painful feet with minimal eye problems. The fifth was a paraplegic veteran from Africa, who could be suffering from human T-lymphotrophic virus I or schistosomiasis (caused by a parasite that is the most common cause of paraplegia in Africa), another one of Dr Gajdusek's favorite research topics. Obviously, this patient did not have Cuban epidemic neuropathy. Only two of Dr Gajdusek's seven patients displayed Sadun's wedge, which was confirmed to be the most reliable clinical marker of the epidemic. In short, the virus was being isolated from the spinal fluid samples *regardless* of the patients' clinical picture.

The virus theory had seemed at one point the likeliest explanation for the Cuban epidemic. Now, in an instant, it was dismissed. Could the viral isolations in so many patients have been the result of simple laboratory contamination?

"It has happened in my lab before—a contaminant that is impossible to find," Dr Gajdusek told his Cuban counterparts. "It happens in every lab doing viral cultures. And it may be impossible to know where it is or what the source of contamination is."

The fact that the Cuban laboratories had previously isolated a Coxsackie A9 virus during a prior outbreak of conjunctivitis made this a convincing option. To resolve the impasse, Dr Gajdusek proposed that we obtain fresh samples of spinal fluid from patients with epidemic neuropathy selected by the Cubans as well as spinal fluid from normal subjects about to receive spinal anesthesia before surgery.

The whole process was to be done in a blinded fashion, with the samples separated in two containers, one prepared by Dr Gajdusek's team and the other by the Cuban virologists. Both samples would be frozen and carried in person to the laboratories under identical conditions. Each laboratory would decide the procedures for viral isolation. If both laboratories obtained independent isolations, then the identity of the virus would be confirmed.

The Cubans accepted the proposal. The American virologists would be leaving in 48 h. The next day, Dr Gajdusek would depart to deliver a lecture in Bilbao, Spain. Fidel Castro gave the order to proceed with the plan immediately.

ENDNOTES

1. Carleton D Gajdusek. CUBA Consultation (op. cit.).
2. Dr Ricardo Santiago-Luis González and neurologist Rafael Estrada, both from the National Institute of Neurology and Neurosurgery in Havana, examined 46 patients with myelopathy at the "Saturnino Lora" hospital in Santiago de Cuba. They concluded that the problem resulted from subacute combined degeneration of the spinal cord due to vitamin B_{12} deficiency.

3. Cuba's newspapers reported on the visit as follows: Fidel visits hospitals in Santiago: "This is the major Cuban effort in the field of health." Dialogue with the press on the effective response to the epidemic neuropathy epidemic. ["Este es el esfuerzo mayor que en salud está haciendo Cuba". Visita Fidel hospitales santiagueros. Dialogó con la prensa acerca de la efectiva respuesta que el país dará a la neuropatía epidémica.] *Juventud Rebelde* (Dominical, 9 de mayo de 1993); E. Palomares Calderón. Fidel provides reassurance, optimism and confidence on the plan to fight the epidemic neuropathy. He sends congratulations to all Cuban mothers. [Reitera Fidel confianza y optimismo en el enfrentamiento a la neuropatía epidémica. Transmitió una felicitación a todas las madres cubanas.] *Trabajadores* (Lunes 10 de mayo de 1993); C. López Gil. Fidel in Santiago de Cuba. "We are using all posible means against the epidemic neuropathy." He visited hospitals where patients with this disease are being treated. [Fidel en Santiago de Cuba. "Estamos empleando todos los recursos posibles contra la neuropatía epidémica". Visitó hospitales donde son atendidos pacientes con esa enfermedad.] *Granma* (Martes 11 de mayo de 1993).
4. CDC: Centers for Disease Control and Prevention. *Ebola (Ebola Virus Disease)* www.cdc.gov/vhf/ebola.

CHAPTER 19

Tobacco Blindness
Isla de la Juventud: May 23, 1993

*The natives smoked perfumed herbs ignited with live coals which
they carried around with them.*
Christopher Columbus

The epidemiologic studies were conclusive: tobacco smokers were five times more likely to develop optic neuropathy than nonsmokers. On May 23, 8 days after first arriving in Cuba, the members of the Mission team flew to Nueva Gerona in the Isla de la Juventud, or Island of Youth (Pine Island), where Dr Eduardo Zacca presented the results of a careful epidemiologic study organized by the Cuban National Institute of Hygiene, Epidemiology, and Microbiology.

Dr Zacca's study attempted to learn why some people had become ill, while others were spared, and what factor or factors were present in some people that increased their risk of developing the disease. Was it possible, too, that there were protective factors in some individuals living on the Island of Youth?

Epidemiologists attempt to answer such questions by using a method called "case-control study," whereby each patient is asked a set of questions aiming to compare the main characteristics of the patient's life with those of a normal control group, matched by age, gender, and neighborhood.[1]

In addition to smoking as a risk factor, the survey unveiled two others that increased the risk of susceptibility to the epidemic disease.

The first one was what the Cubans euphemistically called "irregular food intake." Those who had gone several days without a meal or those who had eaten, at most, only one meal per day were at a higher risk of becoming ill than those who ate regularly.[2]

One of the main reasons for the visit of the Mission team to the Island of Youth was the epidemiologic observation that this island continued to have the lowest number of cases of epidemic neuropathy from the beginning of the outbreak.

Cuban Blindness. DOI: http://dx.doi.org/10.1016/B978-0-12-804083-6.00019-9
135

One of the possible reasons for this remarkable fact became obvious when we learned that the island houses several thousand international students, ranging from high-school students to medical students, as well as students of a teachers' school. The entire educational system (students, teachers, and administrators) in Isla de la Juventud and elsewhere in Cuba was essentially spared the limitations of the *período especial*, and the international students continued to receive a complete balanced diet. This, however, was not true for the rest of the working population on the island among whom isolated cases were reported.

The second risk factor, closely linked to the "irregular food intake" was called "excessive sugar intake." In an effort to alleviate the hunger pains, many individuals had been eating raw sugar by the fistful (*comer azucar a manotadas*), since sugar was the only abundant and inexpensive edible product available on the island.[3]

In the tropics, the cost of eating only sugar is beriberi. Unfamiliar to many, this old scourge used to be an epidemic disease of the nerves and the heart that had been responsible for uncounted deaths in populations that survived mainly on a staple diet of rice and other starches before the discovery of vitamins.

Beriberi had also occurred in Cuba in the 1800s with devastating epidemics among the slaves in sugar plantations. It occurred commonly among the malnourished African and Chinese laborers brought in to work in molasses factories; these poverty-stricken individuals consumed the sugary molasses to complement their measly meals. The disease was known in Cuba as the "edema of the Black and the Chinese" and "molasses disease."

Metabolizing sugars from carbohydrates requires thiamine or vitamin B_1. Since the body is unable to store thiamine, it becomes dependent on the daily ingestion of this vitamin in the diet, mainly through foods such as unmilled rice, bakers' yeast, pork, and eggs.

In the absence of a balanced diet to provide thiamine-rich foods, deficiency of the vitamin develops, and manifestations of deficiency are often triggered by increased consumption of large amounts of sugar or the complex carbohydrates—cassava, yams, plantains, or polished rice—typical of a tropical diet. Beriberi's presence is heralded by burning and swelling of the feet, fatigue, heart failure, and sudden death, as exemplified by José Tomás Solano Echegaray and his fellow inmates at the Ariza prison.[4]

In all likelihood beriberi was present in many cases of the peripheral form of Cuban epidemic neuropathy, particularly among those eating raw sugar by the fistful. However, they rapidly responded to the oral multivitamin pills provide by the government.[5]

But it was the tobacco smoker who most interested the members of the Mission team.

Tobacco smoke contains considerable amounts of cyanide readily absorbed into the body through the lungs. Also, it was well-known among specialists that heavy smokers of the bitter pipe tobacco could, at times, develop a form of blindness called *tobacco amblyopia*, which was quite similar to that observed in Cuba.

At least 74% of the Cuban patients with optic neuropathy were smokers. Therefore, although the clinical matching was not perfect, it was imperative to consider cyanide intoxication from tobacco smoke as one of the possible culprits of the Cuban epidemic.

The Cuban government had already recognized the important link between smoking and the epidemic and since April 1993 had initiated a no-smoking public campaign. In April 1993, an article titled "Quit smoking: it helps" by journalist Carmen Camiñas Lemes appeared in *Juventud Rebelde*.[5] She wrote: "A friend of mine, affected by optic neuropathy, told me that her vision had improved considerably after being admitted for 10 days to the Julio Trigo Hospital in Arroyo Naranjo. This lady, previously a chain-smoker, abruptly quit her toxic habit, and after this, along with a series of medical treatments, including vitamin therapy, her vision improved significantly. She tells me that although not completely cured, at least she has greatly improved…"

The simple secret of the low numbers of optic neuropathy cases in the Island of Youth was now clear: they had regular and balanced nutrition. Upon our return to Havana from the island early in the afternoon, we decided to obtain a telephone consultation with Dr Eric E. Conn, Professor Emeritus in the Department of Biochemistry and Biophysics, University of California at Davis, and one of the world's experts in cyanide-producing plants.[6] I called him from Dr Márquez's desk at the headquarters of the Pan American Health Organization (PAHO)—one of the very few working telephones with international direct dialing in Havana.

Dr Conn confirmed that, in effect, both cassava and tobacco are well-known sources of cyanide; in addition, he also faxed over a list that included a number of tropical plants containing large amounts of cyanogen-releasing substances.[6]

To our surprise, we found many of the Cuban staple foods on the list, including yam, sweet potato, corn, millet, and beans, particularly the small black beans that grow wild in Puerto Rico and Central America. For the past 3 years the staple food in Cuba had been black beans and white rice, popularly called, because of their colorization, *Moros y Cristianos*, "Moors and Christians."

Also, Dr Conn indicated that the variety of lemon grass called *caña santa* was not associated with any well-known poisoning or toxicity, although it was listed as a possible source of mutagens and teratogens of plant origin.

To evaluate the possibility of cyanide intoxication, the Mission team members recommended the inclusion of an expert in neurotoxicity. I proposed the name of Dr Peter S. Spencer, a distinguished neurotoxicologist I had worked with in Colombia and the Seychelles and who had provided invaluable help during my previous research on tropical spastic paraparesis (TSP). After some long-distance telephone calls, his nomination was accepted by the PAHO office in Washington DC.

Dr Spencer was the current Director of the Center for Research in Occupational and Environmental Toxicology at the University of Oregon, Professor of Neurology at the School of Medicine of the University of Oregon for the Health Sciences, as well as Scientific Director of the Center for Environmental Risk Research of the Veterans Administration in Portland, and Consultant for the Environmental Protection Agency. He would be the last member of the Mission to arrive in Cuba.

Dr Spencer was due to land in Havana from Seattle, via Mexico City, 1 week later on May 28. This would create some complications because the members of the Mission to Cuba were scheduled to conclude their research and leave the island on May 25. Therefore, I requested and received permission to extend my stay in order to work with Dr Spencer.

After a concise briefing on the epidemic over the telephone, I asked Dr Spencer to review the neurotoxic causes of blindness involving cyanide, methanol, and other toxins. I also mentioned to him that there were some concerns about possible agents of warfare.

Prior to my departure for Cuba, during my telephone conversation with Dr Frank Young, Rear-Admiral of the Public Health Service in charge of the Office of Emergency Preparedness, he had mentioned the possibility that this epidemic could have resulted from accidental release of Russian-made chemical nerve warfare agents, biologic weapons, or radioactive residues. Most of the Soviet-era military installations had been located at the western end of the island, particularly in the province of

Pinar del Río. For these reasons, to some observers in North America, the East-to-West pattern of increasing prevalence in the distribution of the epidemic, with the largest number of cases in the westernmost province of Pinar del Río, was suspicious. However, no increase in deaths had been recorded on the island during the epidemic and this constituted a major factor against the accidental release of biologic weapons of mass destruction.

Nonetheless, the development of advanced biologic expertise in Cuba, in particular the construction of high-security (Levels 3 and 4) biologic laboratory facilities for handling dangerous agents, was being watched closely in some circles in the United States because of the fear that it could lead to the development of weaponized biologic agents. Nonetheless, and perhaps because of the presence of Dr Gajdusek, the Cubans invited all of the Mission team members to an open visit to their high-security laboratories, including the underground Level 4 military facility in San José de las Lajas near Havana. General Guillermo Rodríguez del Pozo, the military physician in charge of the Neuropathy Task Force, conducted the visit. Apparently, he had been instrumental in the organization of these facilities during Cuba's participation in the African conflicts in order to handle potential occurrence of cases of highly contagious hemorrhagic viral infections such as Ebola or Lassa fever among the Cuban forces. Most of the current research, we were told, was concentrated on the study of human retroviruses, including human immunodeficiency virus, the agent of the acquired immunodeficiency syndrome (AIDS).

★★★

On the night of Sunday, May 23, 1993, two nights prior to departure from Cuba, *Comandante* Fidel Castro Ruz invited all the Mission team members still on the island to attend a formal dinner at the Palace of the Revolution, beginning promptly at 21:30 hours.

Fidel Castro's predecessor, the dictator Fulgencio Batista, had originally intended this building to be the Supreme Court. Following the Revolution, it had been transformed into a splendid palace, brightly illuminated in the tropical night. Inside, the generous use of fine woods and interior green spaces made the rooms inviting and hospitable. Jurassic-age giant ferns from the Sierra Maestra—selected by the late Celia Sánchez, Castro's secretary and confidant of many years—constantly reminded Castro and his visitors of the days of the guerrilla warfare in the mountains.

El Comandante was in an excellent mood. He drank Scotch on-the-rocks, while most of the guests enjoyed Cuban rum *mojitos*. We admired the wood carvings and art pieces in the different salons and then sat down for dinner. The conversation at the table consisted largely of an interesting monologue, as Fidel Castro shared his views of the history of Latin America from Bolivar to San Martín.

The menu was elegant in its simplicity, with a cold seafood appetizer and a main dish of Cornish hens—a gift from a Frenchman who fed the birds a strict diet of *petit pois*—accompanied by small potatoes and rice. Dry white wine from Romania was served. For dessert, we were given *Copelia* vanilla ice cream—which Castro greatly enjoyed, opting for a second serving—and, finally, Cuban coffee. The dinner ended at 2:00 a.m. Back on the street, we could not avoid a guilty feeling of having enjoyed a gourmet meal, while everybody around us was going to bed on an empty stomach.

★★★

The following day, Monday May 24, was scheduled for the closing meeting of the Mission to Cuba, at 2:00 p.m. at the modern Conventions Center in Havana. *Comandante* Castro would be presiding.

Dr Llanos began by thanking the Cuban government on behalf of the Mission for the "open door policy" that had prevailed during the visit and for having provided the members of the Mission with all the information requested. Following these opening remarks, members of the team gave brief presentations in their respective areas of expertise.

The clinical features of the epidemic had been clearly identified: the slowly developing deficits, the tendency to affect both sides of the body in a symmetrical pattern, the beginning of the problems at the most distal points of the toes and fingers, and the very specific location of lesions on very specific *neurons*—*only* the macular color neurons, not the entire retina; only the high-frequency hearing neurons, not those affecting other hearing ranges; and only the largest and longest nerve axons, not the entire nerves—among the billions present in the nervous system.

All these symptoms pointed to the presence of metabolic damage interfering with the production of energy in the cells. In fact, the neurons affected first were high-energy-consuming neuronal groups susceptible to even minor deficits in the production of biochemical energy. This was the hallmark of a nerve toxin or a nutritional deficit.

"We have not been able to find the culprit," I concluded in my remarks, "but we do have a clear idea of its *modus operandi*." The members of the Mission discussed the several factors that could have contributed to the disease. We had finally agreed that the limited diet was the main predisposing factor, combined, in many cases, with the effects of tobacco and alcohol. The virus was considered an irrelevant finding.

Dr Caballero emphasized the need to continue to provide preventive treatment to the entire Cuban population. Since the symptoms had improved in patients treated with vitamins, the plan to provide multivitamins to the entire population was designed, in fact, as a therapeutic proof of the cause of the epidemic. A progressive decline in the number of cases and the eventual disappearance of the epidemic in response to the multivitamins would demonstrate that the cause was nutritional. Dr Caballero also mentioned the importance of improving the supply of basic foodstuffs and recommended more detailed nutritional studies, in particular of the B-group vitamins.

Dr Muci-Mendoza offered help to organize training courses for Cuban ophthalmologists in the new methods of diagnosis successfully used by Dr Sadun. Drs Silva and Thylefors emphasized the need to continue the adequate follow-up and rehabilitation of those patients who had been left legally blind despite treatment. Dr Márquez in closing on behalf of the team indicated the willingness of the Pan American Health Organization/World Health Organization (PAHO/WHO) to continue collaborating with Cuba in eradicating the epidemic.

Then Fidel Castro stood up to close the meeting.[7] He lauded the intensive work performed by the Mission team and made a plea to the international scientific community to continue the research on the Cuban neuropathy: "This is a condition," he said, "that could also occur in other parts of the world. It would be worthwhile that a number of scientific talents in the world would dedicate their efforts to the study of this disease."

Finally, bringing his hands to his beard, his head bowed in a pensive posture, he said to the assembled Mission members, "You have responded to our request for help at this time of need, and we shall not forget you, *compañeros*."

The following day the members of the Mission to Cuba began returning to their home cities: Dr Rafael Muci-Mendoza to Caracas, Dr Juan Carlos Silva to Bogotá, Dr Benjamín Caballero to Baltimore, Dr Guillermo Llanos to Washington, and Dr Björn Thylefors to Geneva.[8] A brief report on the Mission's work would be eventually published.[9]

★★★

It had been decided that I would stay in Havana and wait for Dr Spencer's arrival. Following the departure of my colleagues, I moved from the Biocaribe Hotel and became a guest at the home of Dr Miguel Márquez, the Cuban representative for PAHO/WHO, who enjoyed the rank of ambassador in Havana. Márquez, in his late 50s, was a tall man with a royal girth, curly gray hair always in disarray falling to his neck, a full white beard and a dark mustache framing his easy and friendly smile, and he spoke with the melodious Spanish of the people from the Andes.

Márquez was born in Ecuador in Cuenca, a small Andean city, to which he would eventually return to become Rector of the University prior to joining the WHO. He studied medicine in Ecuador, Chile, Colombia, and Edinburgh, moving from pathology to medical education and finally settling in public health. He had been the PAHO/WHO representative in Nicaragua during the days of the Sandinista victory. In Managua, Márquez learned the difficult art of diplomacy while working with government officials who were former *guerrilleros*—soldiers still redolent of their days in the humid jungles who kept the M-16 rifles under their desks.

In Havana, Márquez was untiring in his efforts working for the common Cubans in the face of the PAHO bureaucracy and the diplomatic corps. He took the epidemic on his shoulders as a personal burden, and much of the international support was due to his behind-the-scenes interventions.

Márquez's wife, Libia Cerezo, had an elegant demeanor and a courtly smile that concealed her maternal tenderness—learned from years of work as a nurse among pregnant women at maternity hospitals in Managua and Havana. Libia was a registered nurse from Cali, Colombia; she had obtained a master's degree in Public Health and had also been a PAHO functionary in Washington DC, where she had met Márquez. Márquez and Libia's had two children, Camilo and Daniela, 12 and 10 years old who Márquez said, were "his fruits of autumn, his dearest prize."

For the first time since my arrival I was able to walk by myself down the narrow streets of old Havana and to wait for the sunset at *el Malecón*. The sea was quiet, and the hues of the setting sun painted in shades of purple the dilapidated façades of the houses lining the avenue. For a brief magical moment, the city felt new again. Night after night a large crowd of citizens filled *el Malecón* to escape the torrid heat of the dark streets and greet the arrival of the cool breezes from the Straits of Florida. Lovers sat facing the sea on damp stones, receiving the blessing of the seawater spray, their silhouettes becoming one in the growing darkness, conversing and caressing—tasting the sea salt in their kisses—oblivious to the noise of the crowd around them. Entire families, grandparents and grandchildren included, walked along *el Malecón* in a spontaneous parade under the moonless sky. Across the bay, barely visible, El Morro Castle stood guard over the dark city.

At the Márquez's home the ambiance was friendly and relaxed. A traditional *Siete Años*, Havana Club rum on-the-rocks, was served before dinner. My hosts and I talked about the events of the day while resting on wicker rocking chairs. Then the 10-hour blackout began, and we had dinner on the patio, among the ferns and the covered birdcages, in the yellow light of gasoline lamps.

Outside, in the darkness, the nightlife of Havana began. Scantily clad *jineteras*, the prostitutes of Quinta Avenida, cast long shadows in the headlights of the few cars venturing into the night. The dark streets, the absence of movement, the quiet houses, the furtive bicycles, the shadows of whispering, faceless people trying to sleep in terraces, balconies, and porticoes, away from the oven-hot bedrooms, everything asserted that this was a city under siege, living at a slow pace through the hardships of an undeclared war.

Date: 5/25/93, Havana, at the Marquez's home. It felt strange to return to a family life after the fast-paced activities of this past week. From Márquez's office I was finally able to call home and talk to my wife Lydia and the boys, Gustavo and Andrés, and to return for a moment to the normal issues of daily life.

The pressures from the 10 hectic days spent with the Mission to Cuba team somehow still linger. Until now my work as a doctor had always taken place in the secluded and intimate space of the examining room, where I built a doctor–patient relationship that allowed me to read the patient's symptoms and find the cause of his or her suffering. Here in Cuba, the epidemic had created an unmanageable number of patients, making the one-on-one relationship barely possible. But, in exchange, case after case had the same features, and you learned to recognize the villain more and more as soon as you approached each patient. Instead of the small audience of students around the hospital bed, this week we did our work in public with WHO and Fidel Castro's government officials and scientists, being followed everywhere by the cameras of newsmen from around the world.

The Mission to Cuba is officially over, but many doubts keep creeping in uninvited. Is malnutrition really the only cause? There has never been an epidemic of this magnitude caused by lack of food. Did we miss something? Is there a critical etiology, a mysterious cause that neither the Cubans nor the Mission could see? The main lingering doubt is the possibility of a widespread neurotoxic poison that could mimic the effects of malnutrition or worsen the effects of vitamin deficiencies. Dr Spencer should be able to clarify these remaining uncertainties.

ENDNOTES

1. Pedro Más Bermejo, et al. [Case-control study of epidemic optic neuropathy, Cuba, 1993] Estudio de casos y controles de la neuropatía óptica epidémica de Cuba, 1993. *Boletín de la Oficina Sanitaria Panamericana* 1995;118 (2):115.
2. Gay J, et al. [Diet factors in epidemic neuropathy in the Island of Youth, Cuba] Factores dietéticos de la neuropatía epidémica en la Isla de la Juventud, Cuba. *Boletín de la Oficina Sanitaria Panamericana* 1994;117 (5):389.
3. Excessive sugar intake with a diet deficient in B-group vitamins, in particular thiamine or vitamin B_1, is a risk factor for beriberi.
4. See Chapter 3.
5. Carmen Camiñas Lemes. [Quit smoking: it helps] Dejar el cigarro ayuda. *Juventud Rebelde* (April 11, 1993).
6. Eric E Conn. Cyanogenic glycosides. *International Review of Biochemistry* 1979;27:21–43; JE Poulton. Cyanogenic compounds in plants and their toxic effects. *Handbook of Natural Toxins* 1983;1:117–157.
7. José A Martin. ["It would be worthwhile that a number of scientific talents in the world would dedicate their efforts to the study of this disease," expressed Commander Fidel Castro during the concluding meeting-summary of the visit of WHO/PAHO's scientific mission. They agree with the hypothesis proposed by Cuba on a multicausal etiology], "Bien vale la pena que un número de talentos en el mundo se dediquen a trabajar en esta enfermedad," expresó el Comandante en Jefe Fidel Castro en el encuentro-resumen de la visita de la misión científica de la OMS/OPS. Coinciden con hipótesis planteada por Cuba en Ginebra sobre el posible origen multicausal. *Granma* (May 25, 1993).
8. José A Martin. [WHO/PAHO's scientific mission departed] Partió misión científica OPS/OMS. *Granma* (May 26, 1993).
9. Mission to Cuba OPS/OMS. Epidemic neuropathy in Cuba. *Boletín Epidemiológico de la Organización Panamerican de la Salud* (OPS) 1992;12:7–10.

CHAPTER 20

Of Poisons and Warfare Toxins
Havana, May 26, 1993

It was inevitable: the scent of bitter almonds always reminded him
of the fate of unrequited love.
Gabriel García Márquez

Dr Peter Spencer was a tall, dapper Englishman. He arrived in Havana, after 2 days of travel from his home in Seattle, impeccably dressed in a khaki safari suit, his starched shirt crisp in the wilting midday heat. He was carrying two leather briefcases, heavy with books. Dr Spencer's renown as a leading expert in neurotoxicology preceded him. Born in London, he had completed his BSc and PhD in Pathology at the University of London and later had come to the United States to work at the Albert Einstein College of Medicine in New York. Although he was a laboratory man by training, his research had taken him around the world in search of human diseases produced by neurotoxins. His laboratory work had consisted of mastering the creation of the "organotypic" nervous system culture, with which researchers recreate in the laboratory environment all the elements that form the peripheral nervous system. A single segment of spinal cord is planted with its dorsal roots, nerves, and muscles, mimicking the normal anatomic structures. This technique would prove to be extremely effective in testing toxic products affecting the nervous system.

Dr Spencer's precise, delicate work and the elegance of his experiments were legendary. He and neurologist Dr Herbert Schaumburg, a pioneer in their field, neurotoxicology, had edited the 3000-page well-known text *Experimental and Clinical Neurotoxicology*.[1] Dr Spencer had done particularly relevant work on nerve lesions produced by clioquinol, the product that was responsible for a large epidemic outbreak, in Japan, of a neurologic condition called SMON, subacute myelooptic neuropathy—literally a slow-developing malady of the spinal cord, optic, and peripheral nerves.[2]

SMON was a mysterious neurologic disease, which made its debut in the 1960s. Before it vanished, the disease had affected at least 10,000

Cuban Blindness. DOI: http://dx.doi.org/10.1016/B978-0-12-804083-6.00020-5

Japanese, some permanently. The symptoms began with abdominal colic and diarrhea, followed by a painful burning sensation in the soles of feet and numbness and tingling in toes and fingers. Half the patients developed paralysis of the legs, and one-third went blind.

Curiously, the number of SMON cases peaked during the months of August and September, every year, from 1965 to 1970. Clusters of SMON cases had been reported, and diarrhea invariably preceded the neurologic symptoms. Thus, an infectious origin was suspected—a waterborne agent, it was theorized, spread by fecal contamination of drinking water. A herpes-group virus, called Inoue–Melnick virus, was isolated but was not conclusively linked to the disease.[3]

Professor Tadao Tsubaki of the Institute of Brain Research in Tokyo uncovered the single most important clue to the mysterious cause of the malady. In 1963, he noticed that some patients with SMON had a greenish discoloration of the tongue. However, not until mid-1970 did scientists from Kagoshima University notice again the green discoloration of the tongue as well as the urine and feces of patients with SMON. Pursuing this lead, the researchers discovered that the greenish discoloration was caused by the presence of iron in a popular drug called clioquinol.[2]

Marketed under the trade names *Entero-Vioform* and *Mexaform*, clioquinol was sold as a gastrointestinal disinfectant and was used worldwide for the treatment of travelers' diarrhea and amebiasis. Among the Japanese, however, the medication had become popular as a diarrhea-*preventive* drug. During the summer months, particularly, the Japanese were consuming clioquinol in large quantities. It was widely believed that clioquinol would not be absorbed by the intestine—until the epidemic disease, with symptoms including blindness, paralysis, and "green tongue," appeared, indicating that the product was circulating in the bloodstream and had reached the nervous system. In 1970, the Japanese government banned the sales of this drug. Within a month, SMON disappeared.

Dr Spencer's familiarity with the SMON epidemic, and the solution that ultimately provided the curative treatment for its sufferers, reinforced the idea that the Cuban epidemic might have a similar toxicologic component.

A dinner reception had been organized at the home of Dr Márquez on May 26 to welcome Dr Spencer. The blackout did not get in the way of an elegant dinner in the candle-lit patio. Minister Teja, General Rodríguez del Pozo, Vice Minister Ramírez, and other important government figures were in attendance. Dessert was being served when, down the dark Havana streets came two black Mercedes limousines flanked by military jeeps. Armed soldiers advanced upon the Márquez home, illuminating it with their flashlights. In their midst, in a circle of light, was *el Comandante* Fidel Castro.

The Cuban leader launched immediately into a learned dialogue with Dr Spencer—a dialogue for which I served as an impromptu translator. Castro felt that the cassava or *yuca* consumed in Cuba was nontoxic. Dr Spencer emphasized that it was a common misconception. It was invariably the case, wherever he had studied toxic diseases stemming from the use or overuse of local food products, that the victims themselves said, "But we have been eating this for generations …"

Castro demonstrated his up-to-date knowledge on the topic of *yuca* in Cuba. The popular "Angolan" cassava, cultivated in Cuba, was not of African origin but of Colombian origin, he said.

Despite his opinion, Castro promised Dr Spencer a meeting the next day at the Cuban Ministry of Agriculture with experts who had studied cassava. Then, after the usual photography session, Castro departed into the darkness of the night.

Dr Spencer noted, after *el Comandante* left, that his soldiers and their rifles had been trained on the foreign visitors throughout their leader's visit.

Early the next morning, Dr Spencer and I had a quiet meeting at the Pan American Health Organization/World Health Organization (PAHO/WHO) headquarters. We discussed the principal question I had asked him before his trip to Cuba: "Could a biological warfare agent be causing the symptoms of the epidemic neuropathy in Cuba?"

In Seattle, Dr Spencer had carefully researched the question. His response was a categorical "No." During World War II, the Nazis were the first to produce nerve gases, which were to be sprayed in a fine mist over enemy lands. Scientists in the Unites States and Russia perfected later versions, and both countries maintain arsenals of nerve gases. Syria and North Korea also have stocks of nerve gases. Sarin, developed in 1938 in Germany as a pesticide and also known as GB, was used during the Iran–Iraq War (1980–1988). Under Saddam Hussein, Iraq twice deployed sarin: in 1983, against Iranian infantry causing tens of thousands of casualties; and in 1988, against a minority Kurdish population killing some 3200–5000 Kurds (www.cdc.gov/niosh/agent/sarin-gb).

Sarin is a clear, colorless, and odorless liquid that evaporates into a gas and spreads easily. Sarin and other nerve agents work efficiently, producing instant death with unfailing accuracy, by using a mechanism akin to that of organophosphate insecticides: they block the transmission of nerve signals in the nervous system. The potency of these agents is appalling. "One tiny drop on the skin of your big toe," said Dr Spencer, "will kill you in 60 s, flat."

There were reports of a recently discovered agent called *Novichok* ("newcomer"), a binary nerve gas secretly developed by the Soviets—8–10 times more powerful than any in the arsenal of the United States. Another new Russian nerve gas was *Agent 33*, probably very close in lethality to Novichok.[4]

However, it was obvious that none of these agents would allow for survival in the event of an accident. There was no evidence that the Soviets had ever transported biological warfare agents to Cuba, and there were no reports of massive mortality among humans or animals in Cuba. On the contrary, no fatalities had occurred among patients with epidemic neuropathy.

Furthermore, the large stockpiles of chemical weapons in the United States and Russia had been slated to be destroyed following ratification of the September 1992 Geneva Convention on Chemical Weapons, which had already been signed by 156 countries.

Next, I told Dr Spencer about our visit to the Level 4 facilities designed for work with extremely contagious and exotic agents. He replied that most of the new hemorrhagic viruses, including Lassa fever virus, Hantavirus, and Ebola (which will become a household name two decades later), originated from Africa.[5] As their name implies, viral hemorrhagic fevers present with fever and hemorrhage in the skin, mucosal and lung bleeding, chest pain, diarrhea, and encephalitis—a clinical picture quite unlike the Cuban cases. Moreover, these agents are highly contagious.

Dr Spencer, with his usual meticulous care and encyclopedic knowledge, took us through a review of other known agents that have been used or proposed as weapons of war: anthrax, botulism, and the pathogens causing plague, smallpox, and tularemia. Smallpox, of course, begins with high fever and sores in the mouth and throat; then a rash develops on the face that spreads to the arms, legs, hands, and feet. The lesions become raised bumps, then fluid-filled pustules, and, after 5 days, scabs.

Anthrax causes fever, malaise, skin ulcers, and fatal pneumonia.

The *Clostridium botulinum* bacteria produces toxins causing botulism, a rapidly progressive paralysis.

Bubonic plague, presents with fever and swollen tender lymph glands—the *bubos*—typically in the armpits or groins. The fatal form produces pneumonia with cough, shortness of breath, chest pain, and rapid death from shock and respiratory failure.

Tularemia, so called because it was first described in Tulare County, California, is produced by infection with a bacterium called *Francisella tularensis*, which first infects rabbits. It presents with fever, skin ulcer, headache, diarrhea, swollen lymph nodes, muscle ache, joint pain, pneumonia, cough, and death by asphyxia.

Unlike the clinical picture of the Cuban patients, all of these agents cause fever, skin lesions, headache, diarrhea, cough, shortness of breath, and muscle ache.

The only infectious agent that produces neurologic paralysis in the absence of fever is botulism, which causes double vision, droopy eyelids, slurred speech, difficulty swallowing, and then weakness of arms, legs, and finally respiratory muscles, producing death by suffocation. The botulism clinical picture is also completely different from that of the epidemic neuropathy in Cuba.

Dr Spencer also mentioned two other chemical agents of mass destruction: chlorine and ricin.

Chlorine gas was used by German troops against France during World War I and recently by Iraq during the Iraq–Iran War. It is converted into acid when it comes in contact with the eyes and lungs, causing cough; burning of eyes, throat, and chest; nausea and vomiting; shortness of breath; and fluid in the lungs.

Poisoning with ricin, which is derived from castor oil, presents with fever, cough, fluid in the lungs, and death.

To complete the picture, Dr Spencer also addressed acute nuclear radiation, which presents with fatigue, lack of appetite, severe diarrhea, and loss of hair, followed by destruction of the bone marrow, as a result of anemia, infection, hemorrhage, and circulatory collapse.

It quickly became clear from the above descriptions that we could eliminate the possibility of a nerve warfare agent or a weaponized microbe as the cause of the Cuban epidemic.

ENDNOTES

1. Peter S Spencer, Herbert H Schaumburg, Albert C Ludolph (ed). *Experimental and clinical neurotoxicology*, Second Edition (New York, 2000).
2. Takasu T, Igata A, Toyokura Y, On the green tongue observed in SMON patients. *Igaku no Ayumi* 72:539 (1970); Igata A, Hasegawa S, Tsuji T, On the green pigment found in SMON patients: Two cases excreting greenish urine. *Jap Med J* 2425:25 (1970); Tsubaki T, Honma Y, Hoshi M, Neurological syndrome associated with clioquinol. *Lancet* i:696–697 (1971); Maede TW, Subacute myelo-optic neuropathy and clioquinol: An epidemiological case-history for diagnosis. *Brit J Prev Med* 29:157–169 (1975); Krinke G, Schaumburg HH, Spencer PS, et al. Clioquinol and 2,5-hexanedione induce different types of distal axonopathy in the dog. *Acta Neuropathologica* 47:213–221 (1979).
3. Inoue YK, Inoue-Melnick virus and associated diseases in man: recent advances. *Progress in Medical Virology* 1991;38:167–179.
4. David Wise. Novichok on trial. The secret of a Soviet nerve gas. *The New York Times* (March 12, 1994).
5. CDC: Centers for Disease Control and Prevention. *Ebola (Ebola Virus Disease)* www.cdc.gov/vhf/ebola.

CHAPTER 21

Three Thousand and Five Hundred Patients Every Week
Havana, May 27, 1993

We follow the principle of distributing what we have among all of us, as in a family.
Fidel Castro

Later in the morning, researchers from Cuba's Ministry of Public Health and from the Department of Medical Geography at the University of Havana briefed Dr Spencer. To date, the epidemiologists told their guest, more than 42,000 cases of the epidemic had been counted, with about 3500 new patients being added every week, and no sign that the epidemic could be abating.

The University of Havana's Dr Luisa Iñiguez—petite, vivacious, with the fine fingers of a pianist—presented all sorts of maps, describing her work with an artistic flair. By careful analyses of case-plot maps she had reduced the geographic points where the epidemic had stricken to their final common denominator. These points, which she called "areales," were usually poor agricultural communities with marginal food production and a history of poverty. In the affected areas of Pinar del Río, for example, substandard nutrition had probably existed for a long time prior to the outbreak of the epidemic. This was an important finding, and its significance would become clear as we tried to explain the peculiar distribution of the epidemic with the highest number of cases in Pinar.

Dr Luisa Iñiguez concluded her presentation with a request: could she play the beautiful Steinway piano that presided over the entrance hall of the World Health Organization/Pan American Health Organization (WHO/PAHO) mansion? "I have not been able to play since the beginning of the epidemic," she said, and then she delighted us by playing with gusto the Cuban composer Ernesto Lecuona's Siboney, La Habanera, Malagueña, and selections from his *Andalucía Suite*.

Early that afternoon, Dr Spencer and I met with the Cuban scientists charged with investigating the toxicologic aspects of the epidemic. At this

Cuban Blindness. DOI: http://dx.doi.org/10.1016/B978-0-12-804083-6.00021-7

meeting, Dr Spencer brought out his copy of *Grant's Toxicology of the Eye*.[1] This massive two-volume work—unavailable in Cuba—listed 51 toxic products known to cause the specific type of visual field defect ("central or cecocentral scotoma") observed in patients with Cuban epidemic neuropathy. Much to our relief, methanol did not produce the typical eye lesion of the Cuban cases. Dr Spencer and the Cuban toxicologists—many in military medical uniforms—methodically reviewed the long list of potential toxins, eliminating agents one by one—from amoproxan (a cardiac drug) to wasp stings; chemicals, such as dinitrochlorobenzene, used in munitions; and the explosive agents di- and trinitrotoluene.

However, Dr Spencer's first question to the Cuban toxicologists was: "Are you completely sure that clioquinol, known also as *Entero-Vioform* and *Mexaform*, is not being used in Cuba as a gastrointestinal disinfectant?"

And, by way of explanation, he added, "I know that it has been banned the world over for 20 years, but sometimes these old-fashioned remedies tend to surface at times when modern medications become unavailable."

Surprisingly, nobody had an answer. After a series of frantic phone calls to the authorities in charge of pharmaceutical products in Cuba, the answer was provided: clioquinol had not been part of the Cuban pharmacopeia since 1972. Dr Spencer's old foe, subacute myelooptic neuropathy of Japan, was, indeed, dead and remained undisturbed in its grave.[2]

After several hours, and numerous rounds of *café Cubano*, the list had been shortened to four possible agents: arsenic, organophosphate (OP) pesticides, trichloroethylene, and chronic cyanide intoxication from tobacco and *yuca*.

Arsenic, the deadly poison favored by mystery writers, became popular in the famous 1940s Frank Capra movie *Arsenic and Old Lace*. Arsenic poisoning is widely believed to have caused the death of Napoleon on the island of Saint Helena, as demonstrated by high arsenic levels in his hair. Dr Spencer was one of the few scientists who embraced my theory that Strachan disease had been caused by arsenic from a source of poisoning in the sugar cane plantations in Jamaica. However, our Cuban toxicologist hosts explained that due to its high toxicity, Paris green (copper acetoarsenite) was no longer used as a weed killer and pesticide in the sugar cane fields and that other arsenic-based products used as ant killers were closely controlled. Once again, and in careful detail, Dr Pérez-Cristiá reviewed the tests for arsenic that had been performed on hair and nail samples from patients with epidemic neuropathy and in healthy controls, showing negative results.

Arsenic, the old killer, had been definitively exonerated in the case of the Cuban epidemic.

One other pesticide remained on the list.

OP pesticides are powerful poisons for both insects and mammals, including humans. These poisons share with the nerve gases used in warfare the deadly ability to block the nerve enzyme cholinesterase. This enzyme is particularly abundant at the microscopic point where the nerve encounters the muscle—the neuromuscular junction. Upon poisoning, all muscles, including those of respiration, cease to function; the poisoned person drowns in his own secretions, stops breathing, and dies from asphyxiation. In nature, the neuromuscular junction is also a favorite target of poisons from venomous creatures, such as cobra snakes, scorpions, spiders, toxic fish and shellfish, frogs, and marine cone-shells.[3] The lethal curare poison of blowgun arrows from the Jíbaro headhunters in the jungles of Ecuador is probably the first example of a cholinergic-blocking warfare agent manufactured by humans.

In the early 1970s, Japanese ophthalmologists noticed an unusually high frequency of eye disorders in people from the Saku region of the Nagano Prefecture, in central Japan, an agricultural area that had heavy pesticide use.[4] It was not until almost 20 years later that the US Environmental Protection Agency (EPA) in Washington DC began studying the ocular effects of OP pesticides and concluded in 1991 that it was necessary to test pesticides for causing possible eye damage before granting license to sell these products.[5]

Although the types of eye problems observed in Saku disease (myopia, or shortsightedness, and retinal or macular degeneration) were completely different from the optic nerve lesion occurring in the Cuban epidemic, as emphasized by Dr Spencer, all potential causes of the epidemic had to be studied. No stone could be left unturned.

The Cubans had already prepared a list of all pesticides and herbicides used in agriculture and those available for domestic use on the island. Using the overhead projector, we began the tedious work of comparing the Cuban products against the long list provided by the EPA and the tables copied from the Grant treatise provided by Dr Spencer.

One by one, Dr Spencer was able to cross out toxic agents from the list, with the exception of methyl parathion, the only pesticide used in Cuba prior to the epidemic. The rest of the toxic agents listed as probable toxins were dismissed without further discussion. Moreover, we were told that from 1990 onward, importation of pesticides from the Soviet Union had ceased. Although this had resulted in tobacco of lesser quality, it also probably resulted in less risk of pesticide intoxication among the Cuban tobacco farmers.

Next, one of Dr Pérez-Cristiá's military collaborators presented the data obtained by the Cuban Neuropathy Task Force. In brief, there had been no universal history of contact with OPs among patients; pesticide exposure in the tobacco-growing areas was similar in both patients and controls; and measurement of acetyl cholinesterase activity in the red blood cells, which is specific for OP exposure, was normal in both groups.

Finally, a young biochemist from the Cuban Tobacco Institute summarized the extensive analyses that had been conducted on Cuban and imported tobacco samples, searching for contaminants, toxins, and pesticide residues in tobacco leaves, finished products (cigars and cigarettes), and combustion fumes. She concluded that no abnormal chemicals or pesticide residues were found.[6]

The sun sets early in the tropics. It was night when we left the Toxicology Institute of Havana. It appeared that OP pesticides (and their deadliest relatives, the nerve gases) were an unlikely cause of the epidemic. The next day we would need to continue the search for two more suspects: trichloroethylene and cyanide from tobacco and *yuca*.

On Dr Spencer's entire list of neurotoxic products was one lone agent known to be capable of causing damage to vision *and* hearing. Trichloroethylene (TCE) is a volatile hydrocarbon used in medicine as an anesthetic gas and in industry as a degreasing agent. In the food industry, it is used as a solvent in the processing of edible oils and decaffeinated coffee.

Strict standards are in place to control industrial utilization and exposure to TCE.[7] Following accidental ingestion or after prolonged industrial exposure, persons intoxicated by TCE suffer loss of vision, deafness, and a peculiar numb feeling at the tip of the nose and in the lips, due to damage of the trigeminal nerves, which carry facial sensation.[8]

In a general sense, TCE intoxication can, therefore, produce the chief manifestations of the Cuban epidemic—blindness, deafness, and loss of sensation. But very few Cubans worked in close contact with TCE or the products in which it is used. What would explain exposure of the entire population of the island? Dr Spencer had done his homework, and he felt that perhaps the answer could be found in the veterinary literature.

A solvent contaminating the soybean meal used to prepare cattle feed produced "Brabant disease of cattle." Described in 1916, it caused hemorrhages in calves and bovines fed soybeans, when oil from the soybeans had been extracted with the solvent TCE.[9] The residual TCE in the soybean meal was toxic to the bone marrow—the place where blood cells are produced—resulting in anemia and hemorrhages due to lack of platelets for coagulation.

In humans, TCE intoxication does not damage the bone marrow but produces lesions of the optic, auditory, and trigeminal nerves. Yet, the significance of Brabant disease of cattle in the context of the Cuban neuropathy epidemic was the possibility that a neurotoxin, such as TCE, could have contaminated the soybean meal. In effect, early in the *período especial*, Cuba had been importing large quantities of soybean meal to provide protein supplements to the population, as an effort to replacing the more traditional proteins of meat, fish, and dairy products in the Cuban diet. Contamination of soybean meal with TCE suddenly seemed to merit careful investigation.

Throughout the morning of Friday, May 28, at a meeting in the boardroom of the Hermanos Ameijeiras Hospital with personnel from the Ministry of Foreign Trade, on behalf of Dr Spencer I discussed the sources of soybean meal consumed in Cuba. I explained that the reason for the urgent meeting was that some of the solvents used to extract soybean oil could be toxic to the nervous system and there was a possibility that these solvents could have contributed to the epidemic.

Foreign trade agents began sending urgent cables to representatives of China, Brazil, Mexico, and Canada, who were the main providers of soybean meal. In each was asked the urgent question: "What is the name of the solvent used to extract soybean oil prior to export of the residual soybean meal?" The hospital offered a frugal lunch while we waited through the afternoon for responses from across the globe. The answer came first from China, followed quickly by answers from Canada, Brazil, and then Mexico. By late afternoon we had all the answers. Each cable bore the same message: TCE was not among the solvents utilized for extraction of oil from soybeans. Why? TCE is far more expensive than other solvents.

Meanwhile, to further reassure ourselves that TCE could be eliminated as a possible contaminant, Dr Spencer suggested I contact Dr David Heikes from the US Food and Drug Administration (FDA) at the Kansas City Laboratory branch. He agreed to analyze samples for TCE and other volatile substances, as well as samples of imported soybean meals and edible soybean oil and sunflower oil distributed to the Cuban population. Months later all tests came back negative.

Although plausible from the epidemiologic and clinical viewpoints, intoxication with TCE was therefore eliminated as a possible cause of the Cuban epidemic neuropathy.

We were no closer to a solution to the mystery than we had been upon Dr Spencer's arrival.

Early next morning, it was now time to turn our attention to the last potential poison—cyanide from *yuca* and from tobacco.

As promised by Castro, Dr Spencer and I had a meeting with the Cuban agricultural experts. The topic was cassava, or *yuca* in Spanish.

For the past several years, Dr Spencer was informed, farmlands across the island had been modified, moving away from sugar cultivation and toward other crops in a drive for agricultural self-sufficiency. Castro himself had encouraged the population to eat more *yuca* and various kinds of yams to reduce dependency on imported wheat.

A new Angolan variety of *yuca* was introduced into the island in 1989. This variety (6329) was obtained by hybridization at the *Centro Internacional para Agricultura Tropical*, International Center for Tropical Agriculture in Cali, Colombia. This variety appeared to be drought resistant, was rich in cyanogenic glycosides, and apparently had a shorter cycle and produced larger and more numerous roots.

Dr Spencer was familiar with the African experience with *konzo* and knew of the dangers associated with cyanide intoxication from improper preparation of cassava. He urged government officials to encourage the citizenry to boil their *yuca* sufficiently, discard the water after cooking, and avoid consumption of spoiled cassava that had been harvested a long time ago or cassava that had been repeatedly frozen and defrosted.

But his warnings were met with amused disbelief. *Yuca* had been part of the Cuban diet for so long that even the scientists greeted his suggestions with a half-smile. How could something as familiar and harmless as *yuca* be blinding the Cuban farmers?

At the meeting, Cuban studies on chronic cyanide poisoning from cassava and tobacco were reviewed. We then learned the reason for the amused disbelief of the Cuban *yuca* experts in response to Dr Spencer's warnings. Dr Hans Rosling from Sweden, the world's leading expert in cassava toxicity, had been invited a year earlier by the Castro government to help the Cubans organize the cassava cultivation effort. Rosling, with his wife and daughter, moved to Pinar del Río, where they lived simply in a typical Cuban house and learned all the traditional Cuban recipes for cooking *yuca*, including the classic *yuca con mojo*. Then, using sophisticated laboratory methods, Rosling demonstrated that human consumption of deep-fried or boiled *yuca* in the Cuban style significantly reduced the levels of cyanide to nontoxic limits.[10]

"There is less cyanide in a pound of *yuca* than in one Cuban cigar," Rosling had concluded.

The compelling conclusion convinced Dr Spencer that there was a low probability that cyanide from *yuca* could have been an important factor in the Cuban neuropathy. Cassava was then released to walk free.

That left, among the remaining viable suspects, only tobacco.

During the next 2 days, Dr Spencer and I visited patients' homes in Pinar del Río, as well as a tobacco factory and a large *tienda*, a cavernous empty government market in Havana. The vacant racks and freezers in the *tiendas* and *bodegas*, and the vanished markets explained the widespread deprivation in the homes of patients—empty freezers and empty kitchen pantries. The poverty and food scarcity everywhere was thoroughly depressing.

Dr Spencer's expertise in electron microscopy of toxic nerve lesions was tapped to review, at the Ameijeiras Hospital with the Cuban chief pathologist, Professor Israel Borrajero, the nerve biopsy samples obtained from patients with different forms of epidemic neuropathy from around the country. Dr Spencer's suspicion was that all of the reported clinical manifestations were expressions of one single disease. All the peripheral nerve biopsy specimens of the sural nerve showed identical lesions regardless of the patient's other complaints or clinical manifestations of the illness.

In every patient, the myelin coverage of the nerve was intact, but the axons inside the nerve were seriously injured. Additionally, there were no signs of infection or abnormal products deposited in these nerves.[11]

The typical lesions of a viral infection—inflammatory cells—were not present, indicating the low likelihood that the damage to the nerve was the result of infection. The observable "axonal neuropathy lesion" appeared to be metabolic in origin—that is, toxic or nutritional factors or a combination of both caused it. However, as the possible toxins were being ruled out one after another, the inescapable conclusion appeared to be that malnutrition was the cause of the epidemic.

Dr Spencer's last day in Cuba was to be spent at the Finlay Institute. The Institute's president, Licenciada Concepción Campa Huergo, was the gracious hostess.

Conchita, as she was known to everyone, was a soft-spoken scientist, whose gentle smile and blue eyes hid a steely determination. She had succeeded quite dramatically—despite the US embargo—in her efforts to place the Cuban-made meningococcal vaccine on the international market. Under her direction the Finlay Institute had developed a research program to produce Cuban-made vaccines against cholera, leptospirosis, typhoid, and rabies, as well as vaccines to protect animals against skin parasites. A visit to the Finlay Institute's factory revealed state-of-the-art equipment, with space-age isolation and air-filtering chambers, in strict adherence to international industrial standards.

Conchita's scientific and administrative merits had been recognized with her appointment as Member of the Politburo. She was one of the most powerful political figures in Cuba. The small group that we met at the modern installations of the Finlay Institute consisted of about half a dozen top-level personnel from the ministries of health, agriculture, tobacco, and toxicology. Dr Spencer and I immediately noted that in contrast to the warm collegiality of all our previous exchanges in Cuba, this meeting was not intended as an open scientific forum for discussion. This time, the politicos would inform us on the official position of the Cuban government on the epidemic. In fact, we were being put on notice that our efforts would no longer be needed. Conchita had found a polite way to thank Dr Spencer and me for the time and effort we had invested in scientific research for the Cubans. This was a farewell meeting.

Outside, in the well-tended gardens of the Finlay Institute, a gentle warm sea breeze brushed across the tall palm trees. Inside the building, in the elegant air-conditioned boardroom, the windows were opaque from freezing condensation.

I knew by heart the figures being presented for the last time. At least 74% of all Cuban patients with optic neuropathy were smokers. Moreover, in comparison with nonsmokers, a history of cigar smoking increased five to seven times the risk of developing visual neuropathy. Also, the risk increased with the number of cigars smoked, resulting in the telltale ascending curve known as the *dose-response effect*: smoking one to three cigars per day multiplied the risk almost nine times. More than four cigars a day, however, increased the risk of optic neuropathy to a level almost 23 times higher than in nonsmokers.

Tobacco smoke contains considerable amounts of cyanide (150–300 mg per cigarette), which is absorbed through the lungs. With a diet poor in animal protein and lacking sufficient sulfur-containing amino acids and B vitamins—in particular vitamin B_{12}, folate, and pyridoxine—failure to detoxify the cyanide from tobacco results in injury to the high-energy neurons in the macula and the optic nerve.

There were other nutritional factors that strongly suggested some toxic connection between tobacco use and diet. The diet of many Cubans was also low in riboflavin and lycopene. The latter is a carotenoid, strong in antioxidant properties, found in tomatoes, guavas, watermelon, and red fruits. Smokers who had appropriate serum levels of lycopene were protected against optic nerve disease, indicating that an oxidative injury could be responsible for the macular damage. It is well known that macular degeneration, a cause of blindness in the elderly, may be prevented by consumption of foods rich in carotenoids.[12]

At the end of several presentations, it had become clear that the party line was to blame tobacco—which, until that day, had been Cuba's boon—as the most important "risk factor" of the epidemic, leaving nutrition to a secondary role. Dr Spencer and I had no questions. Conchita then concluded that the Cuban epidemic was most likely "a toxico-nutritional condition" resulting from tobacco use and lack of vitamins. Appropriate measures had been implemented to control tobacco use and to provide vitamins to all Cubans. According to the government, the problem had been solved, but it would never end for the thousands of victims with neurologic damage, who had permanently lost their ability to enjoy seeing the island's incomparable sunsets and the colorful blossom of the flamboyant trees.

On our way back to Dr Márquez's home, I told Dr Spencer of a remark made by one of the members of the Health Ministry when we were discussing, a week earlier, the proper scientific name for the epidemic: "If you say that the problem is nutritional, then it is the government's fault; if you say it is toxic, then it's *your* fault because you smoke; if we call it 'toxico-nutritional' then we split the blame, but it's firstly *your* fault."

I also remembered a comment from a Havana resident, quoted by a Reuters' reporter: "People have been drinking and smoking in Cuba for years, and they've never suffered anything like this." People knew that the problem was malnutrition. Dr Spencer's visit had led me also to the inescapable conclusion that the cause of the Cuban epidemic—the *primum movens*—was malnutrition. I also had to recognize the Cubans' meritorious efforts to protect, during the worst days of *el período especial*, the groups most vulnerable to the deleterious effects of food scarcity: children, pregnant and lactating women, and the elderly. As Castro had told them: "(We) follow the principle of distributing what we have among all of us, as in a family." Not a single case of epidemic neuropathy had been found among children or women of reproductive age. The elderly, however, often donated their precious food supplements to their grandchildren. Generous and unselfish people like José Polo Portilla then became the first victims of the epidemic in Cuba.

Date: Sunday night, 5/30/1993, Havana, at the Márquez's home: Tonight, the quiet dinner by candlelight in Dr Márquez's patio felt anticlimactic. Perhaps I had known all along that malnutrition was the cause of the epidemic and all the frantic activity of the Mission to Cuba had been a futile effort to deny this reality, trying to find an alternative cause. Officially, the cause of the epidemic was toxico–nutritional. Was that it? Was this the end of my role in the Mission? Would the pressure from the persisting daily patient counts go away? Would I forget the history of privations and hunger? Would I ever find again Sadun's wedge while examining a patient's eyes? Would the nerves and spinal cords of the Cuban patients ever heal?

With an undeniable sense of failure I said goodbye to Libia Márquez and her children and left for the airport with Dr Miguel Márquez. The Mission to Cuba officially concluded on this last day of May. Dr Spencer was returning to Seattle flying from Havana via Mexico, and I was flying to Washington via Miami.

We were leaving the island and saying goodbye to our Cuban hosts without having provided them with a final and conclusive answer as to the cause of the epidemic.

None of the disease models explored up to the final days of the Mission's visit to Cuba explained all the clinical and environmental features observed in these patients. Whatever the cause of the blindness, Dr Spencer and I had confirmed that the culprit did not appear to be arsenic, cyanide, TCE, methanol, or a peculiar pesticide. Neither could the epidemic be explained by any of the strange and exotic ideas that had once tantalized the investigators. It was not some bizarre nerve gas, a new-fangled virus, nuclear radiation injury, or a CIA plot.

The reality was far simpler: Cuba had gambled and placed all its bets on the Communist Soviet Union. With this losing number all its gains had vanished. Its people had been abandoned, penniless and famished. The United States, Cuba's richest neighbor not only refused to provide humanitarian help but also closed all doors by forbidding even other neighbors to provide any assistance to Cuba.

The Torricelli law was working effectively to complicate even the simplest purchase of grain and food. For the foreseeable future, Cubans would continue to survive on rice and beans—supplemented now by a daily pill of multivitamins—and eating brown sugar by the fistful. Smoking and alcohol that had helped ease the hunger pains were being discouraged. The number of cases appeared to have now reached a plateau, at about 4000 per week, but the epidemic still showed no signs of abating.

On the way to the airport, I told Dr Márquez that in my view, there were two main factors against the universal acceptance of the malnutrition hypothesis. One was the absence of previous reports of hunger causing true epidemics of this magnitude with thousands of cases of blindness and peripheral neuropathy; and the second was the Cubans' claim that this was a new disease. Scientists find it hard to accept a disease appearing *de novo* without precedent. It was akin to claiming spontaneous generation—a theory laid to rest in 1859 by Pasteur.[13]

I would eventually find one of the answers to the Cuban epidemic in a 40-year-old monograph written by a Scottish physician about a 1942 outbreak of disease in the far away island of Singapore. But the final answer that had a dramatic link to the political cause of the epidemic would be discovered by Havana's neuroophthalmologist Dr Rosaralis Santiesteban in a century-old report published in 1887 by a forgotten Cuban doctor.

ENDNOTES

1. *Grant's Toxicology of the Eye,* 3rd Ed (Springfield, Ill, 1986).
2. Peter Spencer, et al. (op. cit.).
3. Ninal Senanayake, GC Román, Disorders of neuromuscular transmission due to natural environmental toxins. *J Neurol Sci* 107:1–13 (1992); Gabriel Toro et al. (op. cit.).
4. T Ishikawa, K Ohto, So-called strange disease of Saku, or eye disease due to agricultural agent. *Ganka Rinshoiho* 64:731–733 (1970); T Ishikawa, Chronic optic-neuropathy due to environmental exposure to organosphosphate pesticides (Saku disease). Clinical and experimental study. *Nippon Ganka Gakkai Zasshi* 77:1835–1886 (1973); R Plestina, M Piukovic-Plestina, Effect of anticholinesterase pesticides on the eye and on vision. *CRC-Critical Reviews in Toxicology* 6:1–23 (1978); S Ishikawa, M Miyata, Development of myopia following chronic organophosphate pesticide intoxication: An epidemiological and experimental study. In: *Neurotoxicity of the visual system* (New York, 1980); H Imai, M Miyata, S Uga, S Ishikawa, Retinal degeneration in rats exposed to an organophosphate pesticide (Fenthion). *Exp Res* 30:453–465 (1983); UK Misra, et al., Some observations on the macula of pesticide workers. *Human Toxicol* 4:135–145 (1985); S Imai. A critical evaluation of "The Strange Disease of Saku." *Folia Ophthalmologica Japonica* 1986;37:1351–1354; WK Boyes, Tandon P, Barone S, Padilla S, Effects of

organophosphates on the visual system of rats. *J Appl Toxicol;* Brian Dementi, Ocular effects of organophosphates: A historical perspective on Saku disease. *J Appl Toxicol* 14:119–129 (1994).

5. The EPA Peer Review, dated March 27, 1991 concluded:
"The combined toxicological data from the epidemiological studies and from bioassay demonstrates the potential for organophosphates to produce a wide range of ophthalmological effects, and hence supports the necessity to establish ocular testing as a registration requirement for this class of chemicals for the purpose of hazard characterization and risk assessment."

6. Grupo Operativo Nacional, *Neuropatía Epidémica en Cuba* (Ciudad de La Habana, 30 de Julio de 1993); MINSAP, Dirección Nacional de Epidemiología, Síntesis del estudio epidemiológico en la Provincia de Pinar del Río. (La Habana, 22 de Julio de 1992, pp 1–28.) The OP pesticides purported to produce ocular toxicity included, among others, fenthion, dichlorovos, fenitrothion, malathion, parathion, and methyl parathion. In Cuba, the agricultural products utilized for tobacco included, in order of frequency, mancozeb, zineb, endosulfan, methyl parathion, tamaron, metalaxil, monocrotofos, diazinon, promocarb, and cipermetrina. With the exception of methyl-parathion—widely used around the world—the agents purported to cause ocular toxicity were not used in Cuba.

7. NIOSH, Special Occupational Hazard Review with Control Recommendations: Trichloroethylene (DHEW(NIOSH) Publ 78–130); NIOSH, Criteria for a recommended Standard: Occupational exposure to trichloroethylene (NIOSH-TR-043-73).

8. Defalque RJ. Pharmacology and toxicology of trichloroethylene. A critical review of the world literature. *Clinical Pharmacology and Therapeutics* 1961;2:665–688. Because of the trigeminal anesthesia property, TCE had been used once as treatment for trigeminal neuralgia or tic douloureux; Defalque RJ, The "specific" analgesic effect of trichloroethylene upon the trigeminal nerve. *Anesthesiology* 1961;22:379–384. A rare case of spinal cord involvement has also been reported by Sagawa K et al. Transverse lesion of spinal cord after accidental exposure to trichloroethylene. *Int Arch Arbeitsmed* 1973;31:257–264.

9. Stockman S. Cases of poisoning in cattle by feeding on meal from soya bean after extraction of the oil. *Journal of Comparative Pathology and Therapeutics* 1916; 29:95–107; Pritchard WR et al. Aplastic anemia in cattle associated with ingestion of trichloroethylene-extracted soybean oil meal (Stockman disease, Duren disease, Bravant disease). *Journal of the American Veterinary Medical Association* 1952;121:1–8; Pritchard WR et al. Studies on trichloroethylene-extracted feeds: I-VII. *American Journal of Veterinary Research* 1952;17:425–454. The toxic principle appeared to be a metabolite called S-(dichlorovinyl)-L-cysteine.

10. T Hernández, P Lundkist, L Olivera, R Pérez Cristián, E Rodríguez, H Rosling. The fate in humans of dietary intake of cyanogenic glycosides from roots of sweet cassava consumed in Cuba. *Natural Toxins* 1995;3:114–117.

11. Israel Borrajero, Pérez JL, Domínguez C, Chong A, Coro RM, Rodríguez H, Gómez N, Román GC, Navarro-Román LI. Epidemic neuropathy in Cuba: Morphological characterization of peripheral nerve lesions in sural nerve biopsies. *Journal of the Neurological Sciences* 1994;127:68–76.

12. Seddon JM, Ajani UA, Sperduto RD, et al., Dietary carotenoids, vitamins A, C, and E, and advanced age-related macular degeneration. *JAMA* 1994;272:1413–1420.

13. Spontaneous generation = Pasteur took crude meat, placed it inside a sterilized glass container, and sealed the container using the flame of a Benson burner. The control group he left open, exposed to the microscopic particles floating in the air. After a few days, only the meat exposed to the air showed putrefaction, while the meat in the sealed container remained unspoiled. He demonstrated that putrefaction was caused by bacteria in the air and it was not the results of spontaneous generation.

CHAPTER 22

Changi P.O.W. Camp
Singapore, February 15, 1942

Why are some of the syndromes so prevalent in tropical and subtropical areas but rare in temperate zones when deprivation may be equally or more severe?
Eric K. Cruickshank, MD

Few people landing at Singapore's ultramodern Changi Airport would know that the original runway was constructed with slave labor at the dawn of World War II (WWII). With the two new passenger terminals and the third runway being built on land reclaimed from the Singapore Strait Changi would become one of the largest airports in Asia, securing for the small island nation a solid position as an international air hub in the Pacific rim capable of handling 50 million passengers per year.[1]

Singapore was established in 1819 by Sir Thomas Stamford Raffles as a trading post at the tip of the Malay Peninsula. Along with Hong Kong, the British colonial empire would eventually transform the island of Singapore into a flourishing commercial center and a major naval base. Singapore was called "the Gibraltar of the East" and remained an impregnable fortress of the British Royal Navy until the darkest days of WWII and the battle for control of the Pacific.

On December 8, 1941—a day after the Japanese attack on American naval forces at Pearl Harbor—Lieutenant General Yamashita's 25th Army landed in the narrow neck of Siam's Kra Isthmus. The rapid Japanese invasion of Malaya, Johore, and Singapore was opposed by inexperienced British Army troops who were poorly equipped for jungle warfare and were no match for Yamashita's highly-trained veteran fighters. Yamashita's troops used bicycles to advance fast on jungle trails, pushing the British forces to the southern tip of the Malay Peninsula in only 55 days. After prolonged battles along the Johore strait, on the morning of January 31, 1942, after the last British units had withdrawn into the island, the three-quarter-mile causeway linking Singapore to the Malayan mainland was destroyed.

The siege of Singapore had begun.

Cuban Blindness. DOI: http://dx.doi.org/10.1016/B978-0-12-804083-6.00022-9

Trapped on this small island—26 by 14 miles—were some 52,000 British troops, including the 22nd Indian brigade and the British, Canadian, and Malayan regiments, as well as half a million civilian refugees.

Two weeks later, the British forces surrendered. On February 15, 1942, dressed in khaki shorts, knee-length socks, and shirts with folded sleeves, their heads covered by flat WWI steel helmets, Lieutenant General A.E. Percival, the British Ground Commander in Malaya, and Major General H.G. Bennett, his Australian counterpart, accompanied by their aides carrying the Union Jack and a white flag, surrendered the control of Singapore, once the proudest jewel of the British Empire, to General Yamashita.[2]

The cost of the capture of Singapore to the Japanese was 10,000 dead, wounded, or missing soldiers. They took 70,000 prisoners. About 52,000 Australian and British troops were detained as prisoners-of-war (POWs) in the military prison camp at Changi on the eastern tip of the island.[3]

The fertile tropical land smothered by steaming heat was a laboratory for diseases. Malaria and dysentery, which had been eradicated from the island, became commonplace among the prisoners. Soon, strange deficiency diseases began to occur among the prisoners, aggravated by chronic diarrhea, the scarcity of rations, and the hard manual labor imposed by the Japanese on malnourished soldiers.

Among the military British prisoners at Changi Camp in Singapore was a 28-year-old medical officer, Dr Eric K. Cruickshank. He was a serious-looking physician with a broad forehead, penetrating eyes, and a determined countenance. Born in Dumfries, Scotland, Cruickshank was educated at the University of Aberdeen and at Harvard University, where he was a Thompson Scholar and Fellow in Surgery at the Massachusetts General Hospital in Boston. After the outbreak of WWII, he joined the Royal Army Medical Corps (RAMC) and was mobilized to Singapore in 1940.[4]

During the internment at Changi, as one of the few physicians in the camp, Dr Cruickshank provided medical treatment to over 7000 POWs, as confirmed by his collected notes. Having a full British field hospital in Singapore and a biochemist trained in agricultural research and foods, Dr Cruikshank was able to maintain detailed and accurate records for a period of 3½ years. Dr Cruikshank's Changi records provided one of the few WWII studies, in which the food values could be compared with the pathologic effects of food deprivation.[5]

Dr Cruikshank's most remarkable observations were on previously unknown neurologic diseases associated with poor diets. After the war ended, he reported in detail his first-hand experience with conditions rarely described before, such as "burning feet"—a problem that he himself suffered—as well as beriberi and conditions known as "camp blindness" and "camp deafness."

The most vivid depiction of life at the Changi Camp was written by Ronald Searle in the narrative accompanying his eloquent book *To The Kwai—And Back: War Drawings*. Searle was a young British recruit, who first saw action during the battle of Singapore. He described his arrival at Changi[6]:

> On Tuesday, 17 February 1942, after forty-eight hours of being shoved here and there and awaiting news of our fate, we were disarmed and told that we were to be walked to Changi, an area on the eastern extremity of Singapore Island. Fifteen years before, Changi had been a mixture of uninhabitable mangrove swamp and virgin forest, but it had been since reclaimed as a military base. All troops were to be confined there until otherwise disposed of. In the next two days more than 52,000 of us shuffled the fourteen miles from Singapore in crushing equatorial heat, loaded down with what possessions we had managed to salvage from the shambles of the past few days.
>
> Food was increasingly scarce—a little rice with the odd trimmings, plus anything edible that walked, wriggled or flew, or that could be scrounged or bartered through the wire… Meanwhile clouds of flies clung to us, and an exotic variety of insects ate us. Bites and scratches began to turn into uncontrollable, flesh-consuming tropical ulcers. By summer we were generally ill and ragged… Fortunately, we did not know that this was merely the beginning…

A hospital was organized inside the camp to house up to 2000 sick and wounded men. Surgical, X-ray, and laboratory equipment, as well as medical stores, were transported into the camp. Dr Cruickshank later wrote[7]:

> The hospital was staffed by specialists, medical officers, and RAMC other ranks. Medical and surgical wings were established… Water supply was grossly inadequate for the first four months until repairs had been carried out on damaged pumps and pumping stations. Deep 'bore hole' latrines had to be dug throughout the camp. Electricity was supplied to the camp from September 1942 onwards, sufficient to provide adequate lighting and power for x-ray apparatus and the machines of the workshops… Sea bathing in organized parties was permitted during the greater part of imprisonment.
>
> No issue of clothing was made by the Japanese during the three and a half years' period. The result was that in short time the men's clothes were in tatters and the great majority of outdoor workers wore only a very brief pair of threadbare shorts and wooden Chinese clogs—two pieces of wood covering the soles only and held on by a strap of canvas across the toes. Thus clad the men worked 10 to 12 hours a day in the full tropic sun, many without covering on their heads. Their skins became dark mahogany brown and at a distance they were almost undistinguishable from the Asiatics.

In May 1944, after 2 years of imprisonment, the prisoners were transferred to a prison called Changi Gaol as the Japanese prepared for the demolition of the camp and the construction of the Changi airport—using the prisoners themselves for labor. According to Searle:

> *Changi Gaol squats in a reclaimed swamp on the eastern extremity of Singapore Island… Bleak and positively sinister in the sheer acreage of its stone, concrete and steel, it had been built by the British government only eight years before, in 1936… (It) had been constructed to accommodate six hundred criminals—576 Asian and a mere 24 European. Once we had joined the concentration of prisoners who had been gathered from the surrounding camps, there were over ten thousand of us confined inside… Still more prisoners were eased in by the Japanese, so that by June the gaol itself contained almost ten times more prisoners that it had been designed to hold. Inside the cell blocks it was suffocatingly hot and claustrophobic… the Japanese had allotted one square metre of space to each prisoner… The noise was rarely less than unbearable.*

The deplorable living conditions began to take their toll. Hunger pains, lack of energy, apathy, irritability, and a dejected look soon appeared among prisoners. The steaming humidity of the island kept everyone drenched in sweat, amid the body odors of a crowded humanity. The abdominal cramps and the stench of chronic diarrhea, the fever and the shaking and shivering of malaria, and the unending slave labor heralded the arrival of even more diseases.[8]

Very soon, nutritional problems—resulting from the gross imbalance between the energy expenditure required by the heavy labor and the limited amounts of calories, vitamins, minerals, and the minimum nutritional value provided by the prison camp diet—began to appear. Searle described the prisoners' labor thus:

> *Every day at dawn a party of 900 prisoners was marched off to dig. Fourteen to sixteen hours later they were herded back into the gaol worn out, burnt black by the sun and half blinded by the intense glare thrown up from the sandy ground they were leveling.*
>
> *The Herculean project of flattening the surrounding Changi area and turning it into a military air-base, by using prisoners to reshape the undulating coastal landscape, was well under way. All hills that might be an embarrassment to future runways—and there were several big ones—were to be removed… Hardly surprising that the work finally took one and a half years.*

Cruickshank analyzed the total caloric and nutritional value of the diet, then subtracted the energy expenditure, and correctly concluded that beriberi would be the first disease to appear among the prisoners.[9]

> *The most important deficiency in the diet was that of vitamin B_1. As this vitamin is required for the metabolism of carbohydrates, and a very high proportion of the calories were to come from this source, sufficient vitamin B_1 would be of the*

utmost importance. The early appearance of beriberi was therefore inevitable, and this would probably occur within four months of such diet. Signs of vitamin B$_2$ or riboflavin deficiency might also be expected.

Earlier than anticipated, Cruickshank was proven right:

The earliest clear-cut deficiency syndrome to manifest itself was beriberi. The first cases were seen within the first few weeks of imprisonment but in these there were usually aggravating factors (severe diarrhea, malaria). The first major outbreak occurred from May to July 1942 and the second in late 1944 and during the period of imprisonment in 1945.

Beriberi has been a constant companion of human hunger, occurring always among soldiers and sailors during times of warfare. The Roman geographer Strabo described the disease among legionnaires invading Arabia in 24 BC. The name was coined on the island of Ceylon (Sri Lanka) from the Singhalese word *beri*, meaning weak, emphasizing the great weakness produced by the disease. Descriptions of beriberi are found in the *Neiching*, one of the oldest Chinese medical treatises dating back to 2697 BC, and also in the Japanese text *Senkinho* (640 AD), where it was called "kakke." Epidemic outbreaks of thousands of cases with mortalities over 30% occurred frequently throughout history in the Asian countries where rice was the staple food and in tropical regions with diets based on the complex carbohydrates of plantains and tuberous roots.

The cause of beriberi remained obscure until 1890, when Christian Eijkman, a Dutch physician working in Java, noticed nerve damage in chicks fed only polished rice. The chicks dragged their feet just as his patients with beriberi did. Those who received unpolished rice remained unaffected. This experimental disease called *polyneuritis gallinarum* became a model for the study of beriberi.

In 1911, another Dutch doctor, Casimir Funk, managed to produce an extract of rice polish—the outer husk or pericarp of the grain—which, when given in daily doses to chickens and pigeons otherwise fed on polished rice, would prevent the occurrence of paralysis. Funk used the word "vitamin" to identify this crucial or vital dietary element. The same year in Manila, US Army physician Captain Edward B. Vedder and chemist Robert R. Williams used a purified form of Funk's extract and were able to cure cases of beriberi occurring among Filipino soldiers.

Eventually, in 1939, Williams synthesized thiamine, the anti-beriberi vitamin, facilitating the treatment and particularly the prevention of beriberi. Supplementation of rice, wheat, bread, pasta, and carbohydrate products with pure thiamine virtually relegated beriberi to the museums of medical history.

At Changi, the Japanese fed their prisoners highly polished white rice, and thiamine was unavailable. Soon, Dr Cruickshank had collected notes on 400 cases of beriberi, and more cases kept coming to the hospital. Men complained of numbness, tingling or pins-and-needles in their fingertips and toes, and sometimes the sensation spread to involve the whole body. Rugged infantry men complained of "not feeling the floor" forcing them to walk unsteadily—a finding correlated with the loss of sensation in the joints of the feet and a failure to feel the vibration of a tuning fork on the bony prominences of the ankles and toes. The men returning to camp after a day of labor with aching, stiffness, and tightness or cramps in the calves and a tendency to drag the feet created a sad picture.

But by far the most common manifestation of thiamine deficiency was the swelling of feet noticed by Dr Cruickshank in 317 of his first 400 patients. This was accompanied by breathlessness, lassitude, and weakness in half the cases and by a fast heart rate and cardiac palpitations in some. This was the typical "tropical dropsy" well described by Patrick Manson in his book *Tropical Diseases: A Manual of the Diseases of Warm Climates*, published in 1898 for British physicians serving in the tropical outposts of the Empire.[10]

At Changi, the POWs with beriberi received the only treatments available: bed rest and an English dietary supplement called Marmite—cooked baker's yeast flavored with sugar, salt, and spices. When Marmite was no longer available, the men were given raw rice polishings rich in the missing vitamin. The few remaining ampoules of thiamine were rationed to be reserved for the most severely ill soldiers. Only patients with "kakke," the form of beriberi accompanied by cardiac failure, received intravenous injections of vitamin B_1. As soon as the men returned to hard labor and their prisoners' diet, however, all their symptoms returned.

Some 3 months after imprisonment, a new disease appeared: the men sarcastically called it "happy feet"—probably because it kept them awake all night long. Elsewhere it was popularly known as "burning feet," "painful feet," or "sore feet."

The suffering men were unable to sleep because of the severe burning pain in the soles of the feet. The earliest sign was an intense throbbing in the balls of the feet appearing in the evening at the end of the workday. Some prisoners cut open their shoes or bandaged their feet to alleviate the pain.

Dr Cruickshank, described the burning feet pain in the first 500 cases he observed—as well as the experience of his own suffering, as follows:

> The pains were always worst at night, keeping the patient awake. The patients became worn out from pain and loss of sleep; rapid loss of weight occurred and the appetite often became poor.
>
> Tightly gripping and massaging the feet gave some relief and men adopted a characteristic attitude in bed, sitting forward, cross-legged, gripping their feet. On examination the patient's face wears an expression of chronic distress with dark shadows under the eyes.
>
> The constant pain and the loss of sleep produce an exhausted, red-eyed, irritable patient. Some are almost tearful from the pain. The only abnormal findings in the feet were hyperaesthesia to pin prick and light touch in the majority and excessive sweating in some of the severe cases.
>
> The treatment consisted in the intravenous administration of the di-ethylamide of nicotinic acid resulting in improvement in 68.8% of the cases. In the absence of leafy green vegetables the lack of vitamin B_2 or riboflavin was corrected by giving the patients an extract made from grass and other vegetation that grew around the camp. The leaves were crushed and macerated in water and the extract was drunk. It was effective only when growing green shoots were used.

Then, the blindness hit.

Some 4 months after imprisonment in the Changi POW Camp, Major Orr, an ophthalmologist with the Australian Army Medical Corps, and Dr Cruickshank first began admitting to the hospital patients complaining of failing vision. This would eventually become such a common condition among the POWs that the men called the disease "camp blindness."[11]

In more than half the cases the loss of vision followed an acute attack of dysentery, malaria, dengue, or typhus. Men with obvious signs of undernourishment were more prone to the disease. Prolonged physical effort often precipitated the visual problems. The disease was identical to tobacco amblyopia, but there was no difference in the frequency of occurrence of the disease between smokers and nonsmokers. In some cases the disease appeared 6 to 9 weeks after the diet had improved. The symptoms of camp blindness were described as follows by Dr Cruickshank:

> The earliest symptom was inability to read for any length of time without the eyes becoming tired. This was shortly followed by blurring of vision both for distance and for reading, usually occurring gradually but sometimes suddenly. There might be shimmering or flickering of images.
>
> … there was slight photophobia, and dull aching and boring pain behind the eyes, aggravated by strong light, was common. Some patients were aware of a central blind spot. In patients with symptoms of long duration, pallor of the temporal side of the disc developed, frequently corresponding with the maculopapillary bundle.

Frequently, camp blindness was associated with other diseases resulting from malnutrition, such as painful feet, skin lesions in the genital area, and mouth sores. When the patient was treated early in the course of the disease with Marmite (containing thiamine, niacin, and riboflavin) and with a better diet, there was usually a slow improvement of eyesight.

Many cases, however, continued to progress until signs of irreparable damage to the optic nerve became apparent. Even years after the end of the war, the optic nerves of these patients continued to show the wartime scars. Camp blindness was one of the most frequent problems observed in POW camps during WWII.[12]

Along with blindness came deafness, manifested by a combination of hearing loss and high-pitched noise in the ears, often accompanied by burning feet, loss of vision, and signs of spinal cord damage. It was called "camp deafness," or "camp dizziness" when inner ear vertigo accompanied the hearing loss. The deafness was selective for high-pitched sounds and was compared with an "auditory scotoma."

★★★

Dr Cruickshank was twice mentioned in dispatches for his exceptional medical work in the trying conditions at the prison camp.

After the end of the war, he returned to the University of Aberdeen, where in 1948 he obtained his MD with Honours and the Straits Gold Medal and was appointed to London's Royal College of Physicians.[13]

However, the tropics had become deeply imprinted on Dr Cruickshank's mind, and he could no longer trade the warm perfumed breezes and turquoise seas of a tropical island for the cold and misty air of his native Scotland.

Like Major Plunkett, the mythical hero of poet Derek Walcott's epic *Omeros*, he headed to Jamaica in 1950 to become Professor of Medicine and First Dean of the new Faculty of Medicine of the University College of the West Indies in Mona, Kingston.[14]

Nutritional disorders, specifically their neurologic manifestations, remained his main clinical interest. In 1956, Professor Cruickshank described, for the first time, tropical spastic paraparesis (TSP), a condition he called "a neuropathic syndrome of uncertain origin." This report would open the field for studies on retroviral diseases of the nervous system, which would be undertaken in Jamaica 30 years later.[15]

It was at this time, during the research studies on TSP in Jamaica, that I met Dr Cruickshank's son, Dr Kennedy Cruickshank, who was also interested in TSP. After I had read some of the old publications on nutritional diseases of the nervous system, I called him and he confirmed that, indeed, his father was a survivor of the Changi camp and a leading expert on the neurology of human malnutrition. To my delight, he said that his father was alive and well. He provided his telephone number, and I had the pleasure of talking to Dr Cruickshank Sr on many occasions. At an age when most physicians would have long retired from active work and study, he continued—at 85 years of age—to keep his fingers on the pulse of developments in science and tropical diseases, living in England and Port Antonio, Jamaica.

Upon my request, he sent me, along with his picture, copies of his old typed manuscripts, from which I learned, first hand, of the effects of hunger on the nervous system. Dr Cruickshank's observations demonstrated beyond doubt that malnutrition by itself had produced epidemics affecting thousands of previously healthy young soldiers. The clinical manifestations I had seen in the Cuban patients and which I had committed to my memory, were identical to what Dr Cruickshank had seen in Singapore: beriberi, blindness, deafness, burning feet, myelopathy, suffering, and privations. Even Sadun's wedge—the typical sign of the Cuban optic neuropathy—had been reported by Drs Orr and Cruickshank among POWs more than half a century earlier.

The only missing element now was historical evidence to demonstrate that blindness epidemics had occurred in Cuba before the current outbreak.

ENDNOTES

1. Why are some of the syndromes so prevalent in tropical?: Eric K. Cruickshank. Effects of malnutrition on the central nervous system and the nerves. *Handbook of Clinical Neurology*, vol. 28, Chapter 1 (Amsterdam, 1976).
2. Singapore planning additional expansion of Changi Airport. *Aviation Week & Space Technology* (November 11, 1991).
3. A. Zich and the editors of World War II Time-Life Books. *The Rising Sun* (Alexandria, Virginia, 1977); C. Salmaggi, A. Pallavisini. *2194 Days of War* (New York, 1993); Hiroyuki Agawa. *The Reluctant Admiral: Yamamoto and the Imperial Navy* (Tokyo, 1979).
4. Ronald Searle. *To the Kwai-and Back. War Drawings 1939–1945* (Boston, 1986).
5. Eric K. Cruickshank. *Part One, Camp Conditions and Diet* (Typewritten text, 1946). Copy kindly provided by the author, autographed: "With kind regards to Dr. Gustavo Roman, EK Cruickshank, April 13, 1994."
6. Cruickshank (Op cit. 1946):
   In March 1942, the Japanese issued an official ration scale for prisoners-of-war, which included: rice, 500 g; meat or fish, 50 g; fresh vegetables, 100 g; canned milk, 15 g; flour, 50 g; sugar 20 g; cooking fat, 5 g; the whole with an approximate calorie value of 2360 calories. This theoretical value was never reached, as the rice remained below 400 g for heavy duty rations, for men employed by the Japanese on hard manual labor,

i.e., aerodrome workers, etc. Even these workers received at times only 270 g of rice per day. Light duty workers received often as little as 225 g and those in the "no duty" category, including patients in the hospital, received only 180 g. Soya bean was issued in lieu of rice from December 1943 to March 1944 in quantities up to 170 g per day. Soya beans provide higher calories than rice and are a good source of fat, protein and B-group vitamins. Beans were treated with a cellulose decomposing mold which digested the cellulose and gave the residue a meaty flavor. This preparation, well known by the Javanese, was called "tempi" and was very palatable when fried in oil.

Fish, fresh or dried, was supplied more frequently than meat. Dried fish was a valuable food-stuff. Unfortunately the horse mackerels were not kept sufficiently dried and were delivered in a putrefied condition. At first they were eaten as a necessity with great revulsion, but as the men got used to the "high" flavor they became extremely popular even when in an advanced stage of putrefaction.

After the first few months tapioca, sweet potatoes, yams, pumpkin and cucumber were issued. Green vegetables were late in appearing in the diet in any quantity. Although the soil was poor, agricultural experts in the camp were able to produce an adequate supply of some of the rapidly growing green leaf vegetables—amaranth, Ceylon spinach, and sweet potato tops. Coconut and red palm oils were fortunately supplied in greater quantity than the original scale laid down and the individual issue rarely fell below 20 g per day.

During the first 2 months of imprisonment this diet provided about 2000 calories per day, and as much as 80% of the calories were derived from carbohydrates. The calories in the diet slowly rose thereafter and remained adequate during 1943 and the earlier part of 1944. Thereafter it gradually fell and from February till August 1945 it was grossly inadequate and a state of semi-starvation prevailed. Vitamin B1 was frequently deficient in the diet with the result that vitamin B1 or thiamine deficiency was one of the major medical problems of the camp. Thiamine from highly milled rice, which constituted such a big proportion of the diet, was calculated according to the figures from the Institute of Medical Research at Kuala Lumpur at 0.5 mg/g.

Another important factor which reduced the stamina and physique of the troops was the long hours of manual labor imposed upon men physically unfitted for the task. Owing to the heavy demand for labor by the Japanese, men with chronic diarrhea, frank signs of deficiency disease and immediately convalescent from a malarial attack, were turned out to full duty.

After 2 weeks of imprisonment a careful review of the diet was carried out. The calorie intake was low for members of the Northern European races and the troops could not be expected to go on doing even moderately heavy work on such diet. The protein intake was low. [This probably explained the presence of tropical ulcers, since skin could not be renewed appropriately.] The vitamin A intake was low but reserves would probably last for some time and its deficiency was not likely to occur for some time. Vitamin C was present in the diet in about half the amount recommended by the Technical Commission of the League of Nations. As there was no great storage of vitamin C scurvy might occur.

7. Eric K. Cruickshank, Experiences in the military camp at Singapore. Nutrition of Prisoners of War, *Proceedings of the Nutrition Society* 5:121–127 (1946).

8. This was the same cure that US Army Capt. Edward B. Vedder (*Beriberi*, New York, 1913) had used in 1910, in the Philippines, to cure polyneuritis gallinarum, a form of nerve paralysis produced by feeding chicken a diet of milled rice. Vedder believed—correctly—that this experimental model was identical to human beriberi. See, Robert R. Williams, *Toward the Conquest of Beriberi* (Cambridge, 1961).

9. According to E.K. Cruickshank, Painful feet in prisoners-of-war in the Far East: Review of 500 cases. *Lancet:* ii:369–371 (1946):

   Then sharp shooting stabs of pain in the pads of the toes and in the metatarsals lasting 1/2 to 2s and frequently repeated became superimposed on the dull constant ache. The pains shot longitudinally through the foot and later spread to the ankles and occasionally up to the knees. In some cases they appeared also in the fingers and wrists. As time passed the constant ache became more severe and the sharp pains more frequent.

10. C.M. Fisher, Residual neuropathological changes in Canadians held prisoners-of-war by the Japanese (Strachan's disease). *Canadian Services Medical Journal* 1955;11:157–199.

11. E.K. Cruickshank, Effects of malnutrition on the central nervous system and the nerves. *Handbook of Clinical Neurology* (Amsterdam, 1976). His description of camp blindness, continues as follows:

    On examination there was lowered visual acuity of one or both eyes with difficulty in reading, especially in picking out words or letters in words. Central or paracentral scotomata just above or below the point of fixation were constant findings. They varied in size but were usually larger for red than for black or white. Ophthalmoscopy in the early stages usually revealed a normal fundus. In patients with symptoms of long duration, pallor of the temporal side of the disc developed, frequently corresponding with the maculopapillary bundle.

12. Camp blindness was also seen in many other Japanese civilian internment and POW camps in the Far East between 1942 and 1945. See, for instance: D. Denny-Brown, Neurological conditions resulting from prolonged and severe dietary restriction (Case reports in Prisoners-of-War, and general review). *Medicine (Baltimore)* 1947;26:41–113; W.H. Adolph, et al., Nutritional disorders in Japanese internment camps. *War Medicine (Chicago)* 1944;5:349–355; P.B. Wilkinson, Deficiency disorders in Hong Kong. *Lancet* 1944;2:655–658; P.B. Wilkinson, A. King, Amblyopia due to vitamin deficiency. *Lancet* 1944;1:528–531; J.D. Spillane, G.I. Scott, Obscure neuropathy in the Middle East. Report on 112 cases in prisoners-of-war. *Lancet* 1945;2:261–264; John D. Spillane. *Nutritional Disorders of the Nervous System* (Edinburgh, 1947); C.A. Clarke, I.B. Sneddon, Nutritional neuropathy in prisoners-of-war and internees from Hong Kong. *Lancet* 1946;1:734–737; H.E. Hobbs, F.A. Forbes, Visual defects in prisoners-of-war in the Far East. *Lancet* 1946;2:149–153; D.A. Smith, Nutritional neuropathies in the civilian internment camp Hong Kong. 1942—August 1945. *Brain* 1946;69:209–222.

13. During the Spanish Civil War malnourished civilian populations in Madrid suffered from a similar combination of hearing loss, high-pitched noise in the ears, burning feet, loss of vision and signs of spinal cord damage (F. García Jiménez, F. Grande Covián. Sobre los transtornos carenciales observados en Madrid durante la guerra. *Revista Clínica Española* 1940;4:1–20; M. Peraita, The Madrid Symptomatic Complex— paraesthesia-causalgia syndrome. *Z. Vitaminforsch* 1948;20:1–24; Grande Covián and M. Peraita. *Avitaminosis y Sistema Nervioso.* Miguel Servet, 1a edicion (Madrid, 1941).

14. Derek Walcott. *Omeros* (New York, 1992).

    England seemed to him merely the place of his birth.
    How odd to prefer, over its pastoral sites—
    reasonable leaves shading reasonable earth—

    these loud-mouthed forests on their illiterate heights,
    these springs speaking a dialect that cooled his mind
    more than pastures with castles! To prefer the hush

of a hazed Atlantic worried by the salt wind!
Others could read it as "going back to the bush,"
but harbour after crescent harbour closed his wound.
Omeros Chapter X, Song III, verses 5–7

15. Eric K. Cruickshank. A neuropathic syndrome of uncertain origin. *West Indian Medical Journal* 1956; 5:147–158.

CHAPTER 23

Blockade Amblyopia
Cuba, 1897–1898 and 1991–1993

The loss of vision was called blockade amblyopia because it occurred most often during the four months of the Havana blockade by the American fleet.
Dr Enrique López y Veití (1900)

Dr Rosaralis Santiesteban Freixas had just turned 50 when the epidemic of Cuban blindness began to fill her country's waiting rooms and hospital beds.[1] An elegant woman of European appearance—her last name Santiesteban is common in Madrid and the maternal name Freixas originated from Galicia and northern Portugal—she sported a pageboy hairstyle that had been her trademark since medical school.

Dr Santiesteban graduated in Medicine in 1967 from Havana's Faculty of Medicine, then the only medical school in Cuba. Her teachers, like most academic physicians in Latin America at the time, had trained in France. They inculcated in her the two basic elements—clinical history and examination—that had brought glory to French medicine. Clinical history taking became second nature to her. She carefully questioned and examined her patients looking for the *terrain*—the propitious ground that allowed the disease to germinate. From very early on, Dr Santiesteban had been interested in ophthalmology and neurology, and upon graduation she decided to pursue further studies in both disciplines. She studied at the Pando Ferrer Eye Hospital under the direction of Professor Calixto García, and then she continued her studies in Moscow in the former Soviet Union and later in East Germany—the old German Democratic Republic—where she obtained a PhD; she had a gift for languages, and soon she became fluent in Russian and German.

During the Cuban epidemic, Dr Santiesteban probably examined more patients with blindness than any other eye professional on the island. Her job as Chief of the Department of Neuroophthalmology at

Cuban Blindness. DOI: http://dx.doi.org/10.1016/B978-0-12-804083-6.00023-0

177

Cuba's National Institute for Neurology and Neurosurgery had placed her in early and daily contact with patient after patient complaining of *deslumbramiento*, the painful glare caused by the Cuban sun, and *visión borrosa y ceguera*, blurred vision and blindness. In 1991, when the first cases in Pinar del Río failed to respond to the treatments prescribed locally, Dr Santiesteban traveled to the western province to examine patients herself. Now, 3 years into the epidemic that had swept across her island, she had formed a clear opinion as to its cause.

But holding that opinion, much less expressing it, was tantamount to treason.

Dr Santiesteban was convinced that malnutrition had been ignored as the true cause of the blindness epidemic. The strength to speak out would come from evidence gathered on another island—Singapore—half a world and almost half a century away.

Dr Santiago Luis-González, Director of Neurology at the National Institute of Neurology and Neurosurgery in Havana, showed Dr Santiesteban the notes for his official report to the Government after his trip in April 1993 to examine patients suffering from the myelitis of Santiago de Cuba.

"From the experience of the last few days," Dr Luis-González had written, "it is quite clear that we are facing a funicular myelopathy. This is a nutritional problem. The neuropathy epidemic affects mainly the optic nerves, but the lesion does not stop there. It invades also the peripheral nerves and the spinal cord."

Dr Santiesteban agreed but cautioned her colleague, "This is not the first time that we have provided reports concluding that nutritional deficiencies are the most likely cause of this epidemic outbreak of blindness," she reminded Dr Luis-González. "The Ministry of Health has not even acknowledged receiving those opinions."

Now, though, they might. There was an international scientific mission due to arrive in Havana shortly—the Mission to Cuba composed of representatives from the World Health Organization and the Pan American Health Organization (WHO/PAHO). Their findings, perhaps, would confirm what Dr Santiesteban and her colleagues had been trying to tell their government for over a year—that the true cause of the epidemic of blindness in Cuba was nothing more than chronic, epidemic starvation.

It was not to be. In May, the Mission to Cuba team arrived and began its investigations. Their exhaustive research turned up and then turned aside a half dozen dramatic suspects—everything from toxic pesticides to nerve gases to tainted cooking oils and cyanide in cassava.

In the end, the Mission team had sided with the Cuban Ministry of Health and concluded that the epidemic was probably "toxic-nutritional," a combination of the effects of the unhealthy habit of smoking, drinking, or both, acting upon an organism weakened by a limited and monotonous diet.

Crucially, the international team had strongly recommended that the government provide daily B-group and other vitamin supplements to the entire Cuban population in a simultaneous effort to control the epidemic and to demonstrate its causation by means of this huge therapeutic test. Dr Santiesteban and her colleagues at the Institute felt vindicated for having received at least partial support from the visiting international scientists.

But then veiled criticisms began, suggesting that it was strange that this problem had never been described before in Cuba. These opened the door again to the possibility of wrong diagnoses and inflated figures. Other voices began to suggest that perhaps the falsity was in the description of the epidemic itself. Perhaps the government, weakened by the collapse of the Soviet Union and the continued American embargo, had exaggerated the magnitude of the epidemic simply to garner support from the international community.

The visiting experts had also left the island with the suggestion that this epidemic in Cuba was not a new disease. Similar cases, they said, had been reported in Jamaica during colonial times and in Singapore in prison camps during World War II. Could this be true?

Dr Santiesteban was not convinced. But she remembered hearing as a student an elderly professor mention that during the long wars of independence from Spain, the forefathers of ophthalmology in Cuba had described peculiar lesions of the optic nerves. Dr Santiesteban wondered if those ancient descriptions had ever been published. Probably not, she thought, and she tried to put the matter out of her mind.

In late June 1993, after the Mission to Cuba members had all returned to their home countries, Dr Santiesteban was asked to accompany two ophthalmologists of a Russian mission team visiting the island. After their visit, as a thank-you gift, the visiting doctors presented to Dr Santiesteban a book on the ophthalmologic problems associated with alcoholism. The book was written in Cyrillic, giving Dr Santiesteban a welcome chance to practice her written Russian.

The book also gave her a welcome surprise: it contained a reference to a report published in 1900 by Dr Jose Santos Fernández, a prominent Cuban ophthalmologist from the mid-1800s. His work concerned the toxic effects of alcohol and tobacco in Cuba during the Cuban–Spanish–American war.[2]

This might be the evidence Dr Santiesteban had been seeking. But could she find the 93-year-old book?

Cuba fought three successive wars of independence against its Spanish rulers.[3] The Ten-Year War began in 1868 and devastated Cuba's eastern provinces. It ended with victory for the Spanish and a promise of better representation of Cuba before the Royal Spanish court after the abolition of slavery—although slavery continued until 1886.

Hostilities broke out again in 1879 in the so-called Guerra Chiquita, or "tiny war," that lasted for a year.

Over the following decade, the United States became Cuba's best buyer of sugar and a major investor on the island. At the same time Cuban sugar cane growers and businessmen grew restless under their Spanish landlords. Capitalizing on their restlessness, José Martí, the father of Cuban independence,[4] obtained the rebel forces required to declare war against Spain again on February 24, 1895.

Initially, there was victory for the Cuban rebels. But the Spanish crown was adamant in its desire to maintain control over the island colony. The Spanish forces were placed under the leadership of General Valeriano Weyler.[5] In 1896, soon after taking charge of the forces in Cuba, General Weyler issued his "Reconcentration Camp Decree," placing all Cuban peasants and their families in concentration camps as a way of disrupting popular support for the insurgents. This was an effective measure, but the "scorched land" strategy left the nation's fields uncultivated. Available food supplies were retained for Spanish troops. Thousands of Cubans began to suffer from starvation and from diseases resulting from the appallingly unhealthy conditions of the concentration camps.

By 1898, a year after Weyler's decree, one-third to one-half of the detainees in all the camps had died—as many as 300,000 civilian casualties.[5]

Cuban physicians described their *guajiro* patients as suffering from anemia, weight loss, cachexia, skin pallor, exhaustion, and swelling of the legs, followed by generalized edema and prostration. The victims died suddenly while walking on the streets, as a result of heart damage caused by cardiac beriberi. It was during the terrible year of war, from 1897 to 1898, that the first epidemic of blindness occurred in Cuba.

With a humanitarian tragedy and genocide occurring nearby and American economic interests in the region being threatened, it was not long before the United States government began to take a hand in ending the Cuban conflict.[6] The explosion of the US warship *Maine* in Havana harbor on February 15, 1898, was critical in bringing the United States to the war against Spain.[7] Willing to follow the example of Hawaii, the Americans asked Spain to sell the island to the United States for the sum of US$300 million. On April 20, 1898, President McKinley sent an ultimatum to Spain threatening to intervene in Cuba. In response, on April 21, Spain severed diplomatic relations with the United States. On April 25, Congress declared a state of war between the United States and Spain beginning April 21, 1898. Also on that date a naval blockade of Cuba started. According to David Trask, in his book titled *The War with Spain in 1898*:[8]

On April 22, 1898, (Rear Admiral Sampson's) squadron in double column set a course from Kay West to Havana where he arrived on the same day. Next day he was prepared to close the ports of Matanzas, Cardenas and Mariel and Cabañas. Four days later, Matanzas was closed [because] it has good rail connections to Havana and is well situated to receive supplies from Mexico or the southern Caribbean. The blockade was designed to end all forms of commercial exchange between Spain and Cuba … **The blockade had one serious disadvantage in that it imposed hardships on Cubans**, but its adverse effect on the Spanish troops was considered sufficient compensation for this … At the end of the war, it was revealed that only a few vessels succeeded in running past the American blockade. **Had the war lasted longer the blockade surely would have generated widespread deprivations inside Cuba**. [bold italics added]

The "splendid little war," as Secretary of State John Hay later called it, lasted only 4 months. When it ended, the Spanish Crown ceded to the United States the island territories of the Philippines, Guam, Wake, Wilkes, Peale, and Puerto Rico, in exchange for $20 million. For the first time in its history, the United States was a global imperial power fulfilling what politicians called its *manifest destiny*.

The peace treaty signed in Paris on December 10, 1898, between the United States and Spain also recognized the independence of Cuba under a military U.S. government, resolving a conflict that had lasted 30 years and caused tremendous losses to both Cuba and Spain.

But the conflict had left Cuba in ruins, and, as Dr Santiesteban was about to discover, the final year of the conflict had left many Cubans suffering from malaria, yellow fever, typhoid fever, leprosy, and other ailments, including an epidemic of blindness.

Dr Santiesteban had requested and obtained permission, from the President of the Cuban Academy of Sciences, Rosa Elena Simeón, to conduct a search in the Academy's historical Carlos J. Finlay library. Her quest was finding the scientific publication in the year 1900, quoted by the Russians in their book on alcohol and blindness.

The high ceilings and the dark marble of the walls and columns, in addition to the dim lights, gave the place the air of a medieval cathedral. Old books and papers exposed to the saline humidity of the sea, the constant heat of the tropics, and the lack of sunlight had turned the century-old collections into yellowing, crumbling pages. With a protective surgical mask covering her nose and mouth, Dr Santiesteban began to browse the bookshelves under "C" and soon found the name *Crónica Médico-Quirúrgica de La Habana* engraved in gold letters on the spine of a dark-red leather-bound collection of books. Volume 1 began in May 1875 and Volume 66 ended in March 1940. On page 330 of Volume 26, which had been published in the year 1900, Dr Santiesteban found the article by José Santos Fernández.[9]

Dr Santos Fernández began by describing the increase in cases of blindness that occurred in Cuba during the first Ten-Year War against Spain and then another increase that occurred again during the second and final war of independence.

He quoted a Cuban physician, Dr Domingo Mádan, who, beginning in November 1897, had seen 80 cases of blindness in his practice in the city of Matanzas.[10] His patients also complained of numbness and tingling in the fingertips, feet, and lips, along with a burning sensation in their feet, but the main grievance was the loss of vision.

Mádan's patients had blurred vision, as if there was a veil in front of their eyes. They had difficulty when trying to read or write, threading a needle, or looking at the fine details of their work. There was also a loss of color discrimination: patients confused green and blue, or red and violet, but yellow was identified without problems. Dr Mádan correctly diagnosed this as "central scotoma." Using the ophthalmoscope he saw the typical retinal lesion—what would later be known as Sadun's wedge—and concluded that the lesion was in the optic nerves.

All of Dr Domingo L Mádan's patients had visual loss in a pattern quite similar to that observed in tobacco-alcohol amblyopia—but none of his patients had such "toxic habits." He described the scientific controversy among ophthalmologists who believed that alcohol and tobacco were the main causes of the amblyopia.

However, all of Dr Mádan's patients had endured the penury and privations resulting from the 1896 Concentration Camp Decree, which were compounded by the naval blockade in 1898 imposed by the United States Navy following the declaration of war with Spain.

The similarities of the 1898 report to the current epidemic of Cuban blindness are striking. According to Dr Mádan's original account (translated by Ordúñez Garcia et al.[11]):

All patients presented with the characteristics of central amblyopia. All expressed themselves in the same way: a feeling of fog or cloth that does not allow them to see a familiar face at a certain distance or to read or write; women cannot thread a needle; laborers cannot see the details of their work. … Their visual acuity diminishes substantially and progressively to the point where they cannot distinguish the characters in a Vecker scale either from a short or from a long distance, although they maintain a certain degree of peripheral vision. [The symptoms include,] in some cases, a certain degree of accommodative asthenopia or ocular fatigue [and] dyschromatopsia.[10]

(p 82)

Also, as in the contemporary cases of Cuban epidemic neuropathy, Dr Mádan had found "very characteristic sensory disorders in the limbs":

Almost all patients complain of fatigue or weakness, with pain in the muscle masses. … [T]he feeling of numbness of the toes and fingers is very frequent. … On other occasions they report pain that radiates to the sole of the foot, with hyperesthesia that is described as pain provoked by walking on uneven terrain. The patellar reflexes are preserved in almost all cases, or even exaggerated. … There are no disorders of the orifices or of the area of cranial nerves. There is no amyotrophy.[10]

(pp 83–84)

Dr Santos Fernández had suggested an alcoholic etiology in these cases, but Dr Mádan pointed out that in most of his cases alcohol was not a factor. He suggested a revolutionary idea. The cause of the epidemic of blindness was, in fact, poor nutrition—brought about not only by the rigors of the Cuban wars of independence but also by the American–Spanish war. Mádan wrote:

It is true that our mentor Dr. Santos Fernández has documented in several publications the increase in toxic amblyopia during times of social unrest as the present one. However, the largest number of cases that we have observed are among females from the poor and working class, who live on a small wage, not from the professions that produce [liquor] or that put those women in the situation of unwittingly abusing liquor and spirits. … [A]lmost of them experienced nutritional deficiency and limited economic means.[10]

(p 84)

Citing a previous article by Dr Enrique López y Veitía,[12] titled "Amblyopia due to Malnutrition, Informally Called Amblyopia of the Blockade," Dr Santos Fernández concluded:

> The loss of vision was called 'blockade amblyopia,' because it occurred most often during the four months of the Havana blockade by the American fleet, when the misery had increased due to lack of foreign help in a country already devastated. During this time foodstuffs became so meager that poor families were only able to eat corn, some fruits, sugar, and other less nutritious products.
>
> If there were any doubt about the etiology, that would evaporate if one notices that this pathologic entity so similar to alcoholic amblyopia disappeared from the clinics since the very moment when normality was re-established and its causes ceased."[9]

(pp 333–334)

In the silence of the library, Dr Rosaralis Santiesteban removed her mask and exclaimed: *"Lo encontré!* I found it! This is it!" She had in her hands the proof that similar epidemics of blindness had occurred in Cuba during times of hunger and that the clinical manifestations in 1897–1898 were identical to those found in the current 1991–1993 epidemic. Most strikingly, she found that history had repeated itself and that the epidemic, then and now, resulted from the US blockade of Cuba.[13]

ENDNOTES

1. Rosaralis Santiesteban Freixas. *Epidemias y Endemias de Neuropatía en Cuba* (Ciudad de La Habana, 1997); Santiesteban-Freixas R, Serrano-Verdura C, Gutiérrez-Gil S, Luis-González S, González-Quevedo A, Francisco-Plasencia M, et al. La epidemia de neuropatía en Cuba: ocho años de estudio y seguimiento. *Revista de Neurología (Barcelona) 2000;*31:549–566; Gustavo C Román. La epidemia de neuropatía en Cuba: lecciones aprendidas. *Revista de Neurología (Barcelona) 2000;*31:1–3; Lincoff NS, Odel JG, Hirano M. 'Outbreak' of optic and peripheral neuropathy in Cuba? *JAMA* 1993;270:511–518; Cotton P. Cause of Cuban outbreak neuropathologic puzzle. *JAMA* 1993;270:421–423; Ministry of Public Health of Cuba, Pan American Health Organization, Emory University, CDC Atlanta. International Notes: Epidemic Neuropathy—Cuba 1991–1994. *MMWR (Mortality and Morbidity Weekly Reports)* 1994;43(10):189–192; Llanos G, Asher D, Brown P, Gajdusek DC, Márquez M, Muci-Mendoza R, Román GC, Silva JC, Spencer PS, Thylefors B, Informe de la Misión OPS/OMS a Cuba. Neuropatía Epidémica en Cuba. *Boletín Epidemiológico Organización Panamericana de la Salud* (OPS) 1993 (Julio);14(2):1–4. Coutin-Churchman P. The "Cuban Epidemic Neuropathy" of the 1990s: A glimpse from inside a totalitarian disease. *Surgical Neurology International* 2014;5(1):84.
2. Mills C. In the eye of the Cuban Epidemic Neuropathy storm: Rosaralis Santiesteban MD, PhD. Chief of Neuro-ophthalmology, Neurology and Neurosurgery Institute. *MEDICC* 2011;13(1):10–15.

3. Luis E Aguilar. Cuba, c. 1860–c. 1930. In, Leslie Bethell, editor. *Cuba: A Short History* (Cambridge, 1993); Juan Pan-Montojo (coord.) *Más se perdió en Cuba. España, 1898 y la crisis de fin de siglo* (Madrid, 1998); Hugh Thomas. *Cuba: The Pursuit of Freedom* (New York, 1971).

4. José Martí was the founder of the Cuban Revolutionary Party. Born to middle class Spanish parents in Havana on January 28, 1853, Martí was deported to Spain in 1870, serving a 6-year heavy labor sentence for activities in favor of Cuban independence. He remained in exile in Europe, Mexico, Guatemala, and Venezuela, working as a journalist and writer. In 1881, he went to New York and began to prepare a failed armed invasion of the island, with the support of wealthy Cubans and workers at the tobacco factories in Tampa and other towns in Florida. In 1895, he went to the Dominican Republic and joined forces with Generals Gómez and Maceo, declaring the war for independence on February 24, 1895. Martí died in action, only 3 months later on May 19, 1895, in Dos Ríos. See, Centro de Estudios Martianos. *José Martí, Obras Escogidas* 3 vols (La Habana, 1992); H. Thomas. *Cuba* (op. cit.); Leslie Bethell, editor. *Cuba: A Short History* (Cambridge, 1993).

5. Juan Pro Ruiz. La Política en tiempos del Desastre. In, Juan Pan-Montojo (ed.) *Más se perdió en Cuba. España, 1898 y la crisis de fin de siglo* (Madrid, 1998); Christofer Schmidt-Nowara. Imperio y crisis colonial. In, Juan Pan-Montojo (ed.) (op. cit.); Valeriano Wyler (Wikipedia).

6. John L Offner. *An Unwanted War. The Diplomacy of the United States and Spain over Cuba, 1895–1898* (Chapel Hill, 1992); GJA O'Toole. *The Spanish War. An American Epic—1898* (New York, 1984); David F Trask. *The War with Spain in 1898* (New York, 1981); Lewis L Gould. *The Presidency of William McKinley* (Lawrence, Kansas, 1980); RS Alger. *The Spanish-American War* (Freeport, New York, 1901); H Paul Jeffers. *Colonel Roosevelt. Theodore Roosevelt Goes to War, 1897–1898* (New York, 1996).

7. Hyman G Rickover. *How the Battleship "Maine" Was Destroyed*. With a new foreword by Francis Duncan, Dana M. Wegner, Ib S. Hansen and Robert S. Price (Annapolis, Maryland, 1995); Agustín Remesal. *El Enigma del Maine* (Barcelona, 1998).

8. David F. Trask. *The War with Spain in 1898* (Omaha, Nebraska, 1996) Page 76: In "Synopsis of the War College Plan for Cuban Campaign in a War with Spain" Cpt. Henry C. Taylor, summarized its main points: use of heavy American ships in attacks on the Cuban coast; blockade of the coast and seizure of Cienfuegos and other points as bases; seizure of Bahia Honda or Matanzas as a base for land operations against Havana. Rear Admiral Francis M. Ramsey planned to establish a blockade against Cuba and Puerto Rico. On March 23 Secretary Long issues a detailed plan for the Cuban blockade: The blockage would close the western half of Cuba's north coast, particularly the ports of Matanzas and Havana. Outside the latter port the blockage would consist of a close-in screen including torpedo boats and revenue cutters a second line of protected cruisers, and a third line of armored ships. Some attention was also given to certain ports on the south side of Cuba, particularly Cienfuegos.

    Page 90: The navy had to establish a close blockade of Cuba before attempting offensive operations. This measure would not only exhaust the Spanish army in Cuba depriving it of supplies and reinforcements; it would force the enemy to relieve the beleaguered garrison with naval forces.

9. José Santos Fernández Ambliopía por neuritis periférica debida a autointoxicación de origen intestinal por alimentación defectuosa. *Crónica Médico-Quirúrgica de La Habana* 1900;26:330–334.

10. Domingo Mádan. Notas sobre una forma sensitiva de neuritis periférica. Ambliopía por neuritis óptica retrobulbar. *Crónica Médico Quirúrgica de La Habana* 1898;24:81–86.

11. Ordúñez García PO, Nieto FJ, Espinosa Brito AD, Caballero B. Cuban epidemic neuropathy, 1991 to 1994: History repeats itself a century after the "amblyopia of the blockade" *American Journal of Public Health* 1996;86:738–743; 1997;87: 2053–2054.

12. López E: Ambliopía por desnutrición o ambliopía del bloqueo. *Archivos de la Policlínica* 1900;8(4):85–87.

13. Mádan D. Notas sobre una forma sensitiva de neuritis periférica, ambliopía por neuritis optica retrobulbar. *Crónica Médico-Quirúrgica de La Habana* 1898;24:81–86; Despagnet M. Rapport sur "Note Clinique sur l'amblyopie alcoolique pendant la guerre de Cuba (1868–78)—adressée par le Dr Santos Fernandez de la Havane. *Recueils d'Ophthalmologie* 1891;13:663–665; López Veitía E. Ambliopía por desnutrición o ambliopía del bloqueo. *Archivos de la Policlínica* 1900;8:85–97; Santos Fernández J. Ambliopía por neuritis periférica debida a autointoxicación de origen intestinal por alimentación defectuosa. *Crónica Médico-Quirúrgica de La Habana* 1900;26:330–334; Ordúñez-García PO, Nieto FJ, Espinosa Brito AD, Caballero B. Cuban epidemic neuropathy, 1991 to 1994: History repeats itself a century after the "amblyopia of the blockade" *American Journal of Public Health* 1997;86:738–743 (1996);87:2053–2054; Santiesteban-Freixas R, Pamias-González E, Luis-González RS, Serrano-Verdura C, González-Quevedo A, Alfaro-Capdegelle I, Francisco-Plasencia M, Suárez-Hernández J. Neuropatía epidémica. Proposición y argumentación para renombrar a la enfermedad de Strachan como de Strachan y Madan. [Epidemic neuropathy: proposal and arguments to rename Strachan disease as Strachan and Madan disease] *Revista de Neurología* 1997;25(148):1950–1956; Santiesteban-Freixas R. Historia de la neurooftalmologia en Cuba. *Revista Cubana de Oftalmología* 2005 (Jul–Dec); 18(2); López Espinosa JA. Contribución a la historia de la bibliografía cubana sobre Oftalmología. *ACIMED: Centro Nacional de Informacion de Ciencias Médicas* 2007;15(3).

CHAPTER 24

The End of the Epidemic

Blockade = A restrictive measure designed to obstruct the commerce and communications of an unfriendly nation.
Webster's Dictionary

It was now painfully clear. The epidemic in Cuba neither was a new disease nor was it caused by a secret virus introduced into the island by the CIA. It was not caused by a strange neurotoxin, a warfare agent left behind by the returning Soviet army, the cyanide of cassava, or the smoke from cigars. It was an old disease—perhaps already present among the defenders of Masada and Numantia—a disease as old as warfare and medieval sieges, a disease that had almost vanished from the surface of the earth since the end of World War II when thousands of prisoners of war carrying the scars of malnutrition in the nerves of their feet and eyes and ears were repatriated to the United Kingdom, the United States, Australia, Canada, and New Zealand. The old disease bore a common name. It was called, simply, starvation.

Only a few old doctors working with war veterans remembered the once common "camp blindness" and "camp deafness," and even fewer knew of "happy feet" and other diseases that were spawned in tropical prison camps—diseases induced by the miserable diets and the appalling labor conditions inflicted by the Japanese on their wretched captives.

Most of the survivors from those camps were young men in their early 20s, who returned home prematurely aged when the war ended. Now after decades, they were either dead or dying, and their suffering had been relegated to historical medical records and forgotten. The gripping narratives of medical doctors of the Allied Forces—themselves detained in prison camps in Changi, Hong Kong, the Philippines, Rangoon, Thailand, and Batavia—had made the names of those hellish places synonymous with human suffering and neurologic diseases but had been almost totally forgotten.

Now, half a century later and half a world away, the same stories were repeated by victims of an entirely different war—a war of hunger and

Cuban Blindness. DOI: http://dx.doi.org/10.1016/B978-0-12-804083-6.00024-2

starvation, intensified by the rising and falling political fortunes of two of the world's great superpowers. It was not a cloak-and-dagger American CIA plot or a secret Russian nerve gas. The crisis simply arose from the collapse of the Soviet Union and the continued American embargo against Cuba—combined with the stubborn refusal of Cuba's aging leader to recognize and accept the economic failure of the communist Revolution, which had resulted in the government's inability to properly feed its own citizens.

Dr Rosaralis Santiesteban—flushed with the excitement of her discovery—obtained a photocopy of Juan Santos Fernández's paper. She also followed the historical trail and found in the Finlay Institute library the original article by Domingo Mádan, as well as a copy of his photograph from the eulogy published after his early death at age 40 on July 24, 1898, in Havana; and most importantly, she also located a paper titled *Amblyopia of the Blockade* by Enrique López y Veitía, another Cuban ophthalmologist. Armed with this information, she went to visit Dr Márquez at the Pan American Health Organization (PAHO) building.

Dr Márquez faxed me the articles, and with this new piece of information, I was able to put all the pieces together to solve this epidemic puzzle.

On June 1, 1993, Dr Cruickshank's typewritten pages had revealed a simple truth: the Cuban epidemic was identical in its precipitating cause—the "restricted diet," a British euphemism for hunger—and in its clinical manifestations to the neurologic maladies of World War II prisoners-of-war in Japanese prison camps in the tropics in the Pacific theater of war from 1942 until 1945. Half a century later, every line and every detail of this macabre play was being repeated for an encore in tropical Cuba.

Cuban blindness was also identical to "blockade amblyopia" the epidemic of blindness that occurred in the island among the malnourished pro-independence populations that had survived Valeriano Wyler's concentration camps during the Cuban war for independence only to be further deprived of food during the US Naval blockade of the island during the Spanish–American War of 1898.

Finally, on the first day of June, despite all the political implications, with the coldness of a scientific notation, I wrote in my diary the final entry on the Cuban epidemic of blindness:

Date: 6/1/1993, NIH, Bethesda, Maryland: Epidemic neuropathy of Cuba. Cause: Famine. Until proven otherwise.

AFTERMATH

We are a nation that believes in equality, justice, honesty, and truth.
Jimmy Carter

With the entire data analyzed, all the available evidence pointed to nutritional deficiency as the material culprit in the Cuban epidemic. The deficiencies ranged from protein caloric malnutrition to lack of B-group vitamins, sulfur amino acids, lycopene, and other microelements. The Mission to Cuba had been a purely scientific and humanitarian endeavor. It was beyond our mandate to analyze the complex social, political, and economic factors that had led to this widespread problem of malnutrition.

The underlying nutritional cause of the epidemic was, however, tacitly recognized by the Cuban government, which continued, from 1993 onward, its free distribution of the vitamin supplement Neo-Vitamin II (containing vitamin A and the B-group vitamins—B_1, B_2, B_3, B_6, B_{12}, folic acid, and niacin).[1] And then, on July 15, 1994, Castro himself seemed to admit publicly that it had been starvation that had blinded his people. It was inevitable because, after all, the researchers he had invited from around the world to study the outbreak had agreed on this point: the main causative factor was poor nutrition, more technically called *malnutrition*. However, early in the epidemic, his oppressive rule made it almost impossible for his scientists and doctors to tell the simple truth about the crisis: the failed economy of the Revolution was unable to provide food to all Cubans. The silence imposed on Cuban doctors, or the deliberate deafness to their pleas, cost thousands of Cuban people their eyesight and hearing.

Nonetheless, as eloquently summarized by Ordúñez-García and colleagues: "The Cuban government that [was] the primary cause of the epidemic is the same government that implemented an exemplary health system and that, in the midst of deep economic crisis, was able to protect the most vulnerable population groups from the devastating effects of the recent epidemic, which affected very few children, pregnant women, and elderly people."[2]

In the closing speech of an international workshop on the epidemic of neuropathy, Castro said:[3]

I have the conviction—as an eyewitness in all this process—that vitamins played a major role. This means that we were cognizant that, first and foremost, we had

*to improve nutritional conditions, and we have made, really, extraordinary efforts
in this direction.*

*We are forced to import food from miles and miles away, and transporta-
tion increases the costs, but also, to purchase food we need to use resources and
exchange currency. We had to find new markets for our products.*

*We are excluded from the international monetary and credit institutions, we
have none of that. When we attempt to conduct our own business we are sabo-
taged in the economic arena, to make our life more difficult, in an attempt to force
us to surrender from starvation and disease…*

*If hunger conduces to disease, then this has been a cruel attempt to put our
country on its knees. I prefer not to talk about this, but I feel it is my duty to tell you,
to explain, that when we mention preventive measures, we cannot forget in the
least the food supply of the population; and we are fighting in that front, against
enormous obstacles, but as a first priority.*

The epidemic in Cuba, like all epidemics, showed, in the end, a typical
mountain–peak profile. By May 1993, each and every one of the 11 million citizens of Cuba was receiving the daily vitamin supplement Neo-
Vitamin II. It took one full month before the effects began to be noticed.
In July 1993, the epidemic curve reached its final peak, and from mid-July
1993 onward, it began a steady downward trend. By the end of 1993, the
epidemic had vanished. According to the Centers for Disease Control and
Prevention (CDC), a total of 50,862 Cuban patients had been affected by
the epidemic, making the outbreak arguably the largest documented epidemic of a neurologic disease in the twentieth century.

The rate of recovery with treatment varied widely, depending on the
duration of the symptoms at onset and the severity of the damage. For
instance, half the patients in Pinar del Río were legally blind (visual acuity
20/200 or worse) when the vitamin treatment was started. One year later,
in 70% of them, the vision had improved to 20/40. José Polo Portilla, who
waited too long and was diagnosed initially with untreatable tobacco-
alcohol amblyopia, never improved beyond 20/100. This was enough to
enable him to recognize the faces of his daughter and grandchildren and
to continue working his patch of land, but not enough to read, and he
never regained the color vision to be able to see and enjoy the luminous
red of the Flamboyant trees in bloom in his patio.

Of the 26,446 patients with blindness, about 76% improved substantially, leaving more than 6000 victims legally blind.

How many of the patients with peripheral neuropathy recovered?
According to the Cuban government, "most" or "the majority" of the
24,416 cases improved, although exact figures were never published.

Patients like Pedro Javier, who had severe painful peripheral neuropathy, usually regained or improved their muscle strength and recovered the tactile sensations. The painful burning feet sensation persisted in a minority of cases.

Hearing loss occurred in about 1 in 4 patients with blindness. Deafness was resistant to treatment, and most of these patients were left with a "sound scotoma," which prevented them from hearing high-frequency sounds in both ears.

Treatment with methylcobalamin restored the ability to walk in all patients, like Xiomara, who suffered from myelopathy due to vitamin B_{12} deficiency. Most—also like Xiomara—regained bladder control. But she, however, remained legally blind, and her graceful gait was also lost. She walked very stiffly, using a walker at home or a cane when going out.

The recommendations of the Mission to Cuba were carried out, and a wave of international solidarity responded to the appeal from the WHO's Disaster Relief Program to help Cuba in this moment of need. The European Community, in particular Cambridge University in the United Kingdom, offered research support and short-term training visits to Cuban scientists. Sweden sent Dr Hans Rosling and Dr Per-Anders Lundquist to study specifically the problems related to cassava consumption in the Cuban epidemic. Spain, Italy, Russia, and other countries also sent scientific missions.

In contrast, the US government rejected numerous pleas from the medical community to lift the embargo on food and medicines for humanitarian reasons. Moreover, in view of the unexpected strength of the Castro Government in dealing with the Torricelli law, the 104th US Congress approved the Helms-Burton Act—also known as the Cuban Liberty and Democratic Solidarity (Libertad) Act of 1996—intended to continue the blockade. Its stated intent was to "seek international sanctions against the Castro government in Cuba, [and] to plan for support of a transition government leading to a democratically elected government in Cuba." The Helms-Burton Act received almost universal condemnation from the international community of nations, since it ran counter to the spirit of international law and national sovereignty.[4,5,6,7,8]

Despite the troubled political relationship between the two nations, exchange of medical information continued. Cuban Health Vice Minister Antelo, accompanied by Dr Asher and me, as well as by USPHS Rear-Admiral Gary R. Noble, CDC Associate Director, Washington, visited Dr Carleton Gajdusek's high-security Level 4 laboratories in Fort Detrick,

Maryland, to observe animal studies being conducted with samples from Cuba. Later on, Dr Pedro Mas from the Pedro Kourí Tropical Medicine Institute and Dr Pilar Rodríguez from the Center of Genetic Engineering and Biotechnology traveled to Dr Gajdusek's laboratory at the National Institutes of Health (NIH) in Bethesda, Maryland, to continue the virologic research. However, confirmation of a causative virus by US laboratories was never achieved.

Dr Benjamin Caballero returned to Cuba to provide advice to the National Institute of Nutrition. A second delegation from the CUBA–USA project also visited the island in June 1993. Vice Minister Abelardo Ramírez, with Dr Jorge Haddad and Dr Daniel Rodríguez Milord, visited the CDC and the NIH to update the epidemiologic surveillance network in Cuba.

Evaluation by Cuban nutritionists confirmed the lack of vitamins and sulfur-amino acids in the diet of Cubans. A study at a workers' dining facility in Pinar del Río during the second semester of 1991 showed the following percentages of US Required Daily Allowances: vitamin A 33%, thiamine 50%, vitamin B_6 57%, B_{12} 62%, and folic acid 29%.

Epidemiologists confirmed the near-absence of cases of neuropathy in Cuba among children and people younger than 15 years of age and among pregnant or lactating women. Traditionally, these are the groups most seriously affected by neurologic diseases of nutritional cause in the tropics—for example, the majority of African *konzo* cases. However, the fact that, despite the dietary restrictions, these groups continued to receive nutritional supplementation, including dairy products and vitamin supplements for pregnant women, probably protected them from being affected by the epidemic disease.

The Cuban epidemiologists also confirmed that less-than-regular meals increased the risk, and they discovered that protection was afforded by intake of protein, animal fat, methionine, and the group of B-complex vitamins, notably vitamin B_{12}, riboflavin, niacin, and pyridoxine.

Interestingly, people who raised one or two hens at home for eggs were fully protected, as were those who had helpful relatives overseas (an indication of the strength of the US dollar in the Cuban black market even during the darkest hours of *el período especial*). Also, no association was found between consumption of home-brewed alcoholic beverages and an increase in risk.

The most important study on the Cuban epidemic would be undertaken by the Atlanta-based CDC, coordinated by Drs Rossanne Philen and Caryn Bern. Dr Philen was a Medical Epidemiologist from the CDC's Health Studies Branch, National Center for Environmental Health, and Dr Caryn Bern was a Medical Epidemiologist from the CDC's Division of Nutrition, National Center for Chronic Disease Prevention and Health Promotion. I returned to the island for this study and served as the liaison between the American and Cuban teams.

The US–Cuba study was conducted in September 1993 in Pinar del Río. This probably has been one of the largest cooperative efforts ever conducted in Cuba by members of the US Federal Government, including the CDC, the US Food and Drug Administration (FDA), and the NIH, as well as by Emory University, the Cuban Ministry of Health, and the Pan-American Health Organization (PAHO).

Hundreds of Cuban family physicians, dietitians, nurses, and supporting personnel participated in the study, which involved visiting the homes of 300 persons, half of them patients with ocular damage and half of them normal control subjects matched by age, gender, and municipality of residence.

Every person selected for the study was admitted to the Abel Santamaría Hospital for a 1-day medical, ophthalmologic, and neurologic examination using state-of-the art instruments and procedures, including photography of the retina. Hundreds of blood and urine samples were obtained and transported to the United States for analysis. Thousands of pieces of information about diet, occupation, activity, and environmental exposure were analyzed.

The results were eventually published in the prestigious *New England Journal of Medicine*. The study concluded: "The epidemic of optic and peripheral neuropathy in Cuba between 1991 and 1993 appears to be linked to reduced nutrient intake caused by the country's deteriorating economic situation and the high prevalence of tobacco use."

It is plausible that Fidel Castro—who closely watched the course of the epidemic, received daily briefings on the work of the Mission to Cuba and quickly recognized its cause—rapidly understood that it was impossible to make any progress without dealing with the malnutrition in his people. Despite the soybean supplements and the expensive importation of foodstuffs, the situation reached a point of no return for the inefficient Soviet-like state farms. Fidel Castro tacitly acknowledged that he could not be a tyrant and a generous father at the same time, as dreamed by Stalin. On July 31, 2006, he left the office of President of Cuba and was succeeded by his brother Raúl Castro.

Perhaps as a result of his growing awareness of the economic realities, within 3 months of the conclusion of the visit of the Mission to Cuba, Fidel Castro's government implemented a number of economic measures—many of them unheard of before on the Communist island. From that moment on, Cuban people would need to depend more on their own initiative.

First, in August, circulation of the US dollar became legal. One month later, authorization for self-employment and private industries and businesses was granted (*trabajo por cuenta propia*), and new agricultural and farming cooperative ventures replaced state-run farms.

Almost by a miracle, food became plentiful again; the farm markets were offering fresh fruits and vegetables. Small family-owned restaurants—called *paladares* (palates) for their tasty food—opened across Havana.

These measures were then followed, in 1994, by reduction of the military budget, price increases, and reduction of state services previously provided free. Finally, late in the year, a new Tributary System Law implemented income and property taxes nationwide.

Heavy economic pressure was brought on tobacco and alcohol, the two largest risk factors in the epidemic, other than diet. As of June 1994, the box of "Popular" cigarettes increased from 30 cents to 2 pesos for those on the ration list, but to 10 pesos for everyone else. A bottle of domestic rum climbed from 3 to 40 pesos, and a bottle of beer from 25 cents to 1.20 pesos. With the improvement of food supplies resulting from economic measures, a box of 30 tablets of Neo-Vitamin II was sold at 1.50 pesos. Last, the state ended the era of no-cost, free-attendance of sport, artistic, or cultural events: a clear sign was given that Cuba had entered the global market economy. From then on, the word "gratis" disappeared from the island.

The epidemic also brought Cuba out of its isolation.

The initial chords of "Chan-Chan" played on the Cuban *tres* by Compay Segundo filled the concert hall and were immediately recognized by the delighted cheering crowd attending the first European screening of the documentary film *Buena Vista Social Club* in Amsterdam in 1998. The success of what would have been otherwise an old-fashioned geriatric orchestra—Compay Segundo died at 95, Rubén González at 84, and Ibrahim Ferrer at 78—can only be explained by the Cubans' rapport with the audience, the true artists' capacity to express in the soulful *boleros* the universal tenderness and sadness of unrequited love, or the *alegría*, the joie-de-vivre of the *son montunos* from long ago. The scenes from Director

Wim Wenders' movie document the joyful, emotional response of the Dutch public to the rhythmic songs from the far-away tropical island, performed by a group resurrected more than 50 years after the original club in Havana had closed. American musician Ry Cooder, who organized the group, said that he "went searching for the sounds of an island and discovered the soul of a people."

Finally, Cuba showed that it was more than sugar and socialism. The epidemic had changed forever Cuba's image. The music inspired a truly international passion for Cuba and brought along a fashion for *mojitos*, *café cubano*, and Cuban *pernil* sandwiches. Travel to the island boomed. Spain first, and then Italy, the Netherlands, the rest of Europe, Canada, Mexico, and other Latin American countries began offering vacation tours to the unspoiled beaches and underwater reefs of Cuba. A new industry was born on the island, with instant economic results. Cuba's GDP increased by 6.2% in 1999, with tourism surpassing sugar as the primary source of foreign currency. In that year alone, 1.6 million tourists visited Cuba, resulting in US$1.9 billion in gross revenues. The numbers doubled a decade later, and currently tourism in Cuba is an industry with over 3 million arrivals per year. With the normalization of diplomatic relations between the United States and Cuba in 2015, the number of American tourists to the island will continue to increase.

In the final analysis, ironically, it took a devastating event like the neuropathy epidemic for Cuba to be propelled out of the shadow of Soviet socialism into the bright international community of nations.

San Antonio, Texas, 2009—Houston, Texas, 2015

ENDNOTES

1. [Neo-Vitamin II is born] Nace el Neo-Vitamin II. *Granma* (15 de Abril de 1993); [Vitamins for everyone] Vitaminas para todos. *Tribuna de la Habana* (Domingo 25 de Abril de 1993); [Information on Neo-Vitamin II] Información sobre el Neo-Vitamin II. *Granma* (28 de Abril de 1993).
2. CDC: International notes. Epidemic neuropathy–Cuba, 1991–1994. *Morbidity and Mortality Weekly Report MMWR* 43:183–192 (March 18, 1994).
3. Miguel A. Márquez. [Report of WHO/PAHO on control, treatment, research and epidemiological surveillance of epidemic neuropathy] Informe de la cooperación OPS/OMS en el control, tratamiento, investigación y vigilancia epidemiológica de la neuropatía epidémica. In, F. Rojas Ochoa (ed.) (Op cit. 1995); Claude de Ville de Goyet, MD, Director Emergency Preparedness and Disaster Relief Coordination Program, PAHO/WHO (Havana, 25–28 May 1993); PAHO/WHO. *The Epidemic Neuromyelopathy in Cuba: An Appeal to the International Community for Humanitarian Assistance* (Washington, June 1993).

4. List of Scientific Missions. *European Community Delegates:* Mr Eduardo Lechuga, EC representative for Latin America, Mexico, Chief of the Mission; Dr Fernando Martínez, Director of the Carlos III National Epidemiology Center, Madrid, Spain; Prof. P.K. Thomas, Neurologist, London UK; Dr Peter Baxter, Epidemiologist-Toxicologist, Department of Community Medicine, Cambridge University, UK; Dr Nikola Schinaia, Epidemiologist, National Institute of Tropical Medicine, Rome, Italy; Dr Isabel Barrientos, Regional delegate for Latin America, International Red Cross and Half Moon, San José, Costa Rica; Dr Guillermo Llanos, PAHO, Washington, DC; Dr José Luis Ceballos, PAHO, Washington, DC (Havana, 16–21 June 1993). Delegation from the UK: Dr C.J. Bates, Nutritionist, MRC, Division of Nutrition, Cambridge University; Dr J.P. Baxter, Toxicologist, Cambridge University; Dr R. Plant, Neuroophthalmologist, National Hospital, Queen's Square, London; Dr M.G.M. Rowland, Epidemiologist, PHLS Communicable Diseases Surveillance Center, London; Prof. P.K. Thomas, Neurologist, University of London (Havana, 15–26 August 1993). Spain: Delegation from the Ministerio de Sanidad y Consumo de España: Dr Ignacio Abaitua, Internal Medicine, and Dr Eduardo Rodríguez Farre, Toxicologist. Italy: Dr Luigi Cafiero, Consultant for Humanitarian Affairs and Cooperation, Rome, Italy. Russian delegates: Dr Maleev Victor Sergueevich, Epidemiologist, Ministry of Health; Dr Stepanenko Alexev Alexevich, Neurologist, Institute Semashko; Dr Lysenko Vera Sergueevna, Ophthalmologist, Ministry of Health; Dr Savinov Alexandr Petrovich, Virologist, Institute Semashko; Dr Yartev Mijail Nikolayvidi, Immunologist, Institute Semashko (Havana, 29 June-27 July, 1993). CUBA-USA delegation: Mrs Mary Murray, Coordinator; Mrs Joanna Cagan; Dr Katherine Tuker, Epidemiologist; Dr Jean Handy, Virologist, University of North Carolina; Dr Thomas R. Hedges, Neuroophthalmologist, Tuffs University; Dr Samuel Sokol, Ophthalmologist and Clinical Neurophysiologist (Havana, 28 June–5 July, 1993).
5. J Gay, C Porrata, M Hernández, et al. Factores dietéticos de la neuropatía epidémica en la isla de la Juventud, Cuba. *Boletín de la Oficina Sanitaria Panamericana* 117:389–399 (1994).
6. Members of the CDC involved in the Pinar del Río study included: Dr Gary R. Noble, CDC Associate Director/Washington, Assistant Surgeon General USPHS; Dr Joe H. Davis, CDC Director, International Health Program Office, Assistant Surgeon General USPHS, Associate Director for International Health; Dr Edwin M. Kilbourne, Assistant Director for Science, Epidemiology Program Office; Dayton Miller, Mark A. Miller and Josephine Malilay, CDC Medical Epidemiologists, in addition to Drs Philen and Bern. For the FDA: William R., Obermeyer, PhD, Research Chemist, Division of Natural Products, Biological and Organic Chemistry Branch, Food and Drug Administration, Washington. For the NIH: Gustavo C. Román, MD, Chief, Neuroepidemiology Branch, National Institute of Neurological Disorders and Stroke, National Institutes of Health, Bethesda, MD. For Emory University: Nancy J. Newman, MD, Neuro-Ophthalmology and Neurological Diseases, The Emory Clinic, Atlanta; Frederic E. Gerr, MD, Division of Environmental and Occupational Health, Emory University School of Public Health. For Cuba: Neurology, Drs Rafael Estrada, Nelson Gómez, Antonio García, Marta María de la Portilla; Internal Medicine, Dr Carmen Serrano; Ophthalmology, Dr Melba Márquez, Laila Elena Teira, Blanca Emilia Elliot, Carlos Perea; Epidemiology, Mariluz Rodríguez; Neurophysiology, Dr María C. Pérez; Administration, Vice Minister Abelardo Ramírez, Drs Jorge Haddad and Rodríguez Milord. For PAHO/Cuba: Dr Miguel Márquez, Dr Julio Suárez and Dr Montalvo.

7. The Cuba Neuropathy Field Investigation Team, Epidemic optic neuropathy in Cuba—Clinical characterization and risk factors. *New England Journal of Medicine* 333:1176–1182 (1995).

8. Alfonso P: Panorama de las reformas económicas en Cuba: 1993–1994 (*The Miami Herald*). In, *Cuba en Transición*, Volume 4. See the complete text at: http://www.lanic.utexas.edu/la/cb/cuba/asce/cuba4/alfonso.html. Castañeda RH, Montalván GP: Cuba 1990–1994: Political Intransigence versus Economic Reform Inter-American Development Bank. In, *Cuba en Transición*, Volume 4. See complete text at: http://www.lanic.utexas.edu/la/cb/cuba/asce/cuba4/castaneda1.html.

INDEX

Printed in the United States
By Bookmasters